GOODS OF THE MIND, LLC

SAT Targeted Practice
Math

SAT Reasoning Targeted Practice Series

Cleo Borac, M. Sc.
Silviu Borac, Ph. D.

This edition published in 2013 in the United States of America.

Editing and proofreading: David Borac, B.Mus.
Technical support: Andrei T. Borac, B.A., PBK

Send all inquiries to:

Goods of the Mind, LLC
1138 Grand Teton Dr.
Pacifica
CA, 94044

SAT Reasoning Targeted Practice
SAT Targeted Practice - Math

Contents

What this Workbook Can Do For You

This book is for students who want to excel in the American academic landscape. While many people think the SATs are a one-time hurdle, we hold the view that the PSATs and the SATs are by no means the end, but the beginning of a series of tests: GRE, GMAT, MCAT, LSAT, and other graduate level tests. By working on the SAT top-down - that is, developing the skills needed to solve all the problems rapidly from scratch - you will be much better prepared for the graduate level standardized tests. We want to help students put their effort where it can do the most for them *in the long run*. Solid knowledge and the skill to figure things out from scratch will never go out of fashion!

Our philosophy is completely opposite from that of many popular "quick fix" training methods. Most students prefer to be taught "tips and tricks" about how to select a correct answer without actually solving but by exploiting every possible bypass: plugging in answers, eliminating choices, and using elaborate guessing schemes. The truth is, on their own, these half-baked methods *max out at around 680 points* (per section). Why? Because the test writers *always* include questions that cannot be solved by these methods. After all, the purpose of the test is to select students with superior academic potential, and the writers of the SAT have the freedom to adapt the problems to fulfill this goal. The many reasons we ask students to make the effort to learn problem solving include:

- to be sure and safe on the problems with unexpected twists

- to be able to obtain a high score and not max out under 700

- to acquire lifelong problem solving skills that will help on graduate level tests as well as during employment

9

- to develop strong general test taking skills that will serve them well on any future test

- to improve their problem solving speed by designing efficient strategies even if one specific problem can be solved by plugging in answer choices

The SAT is not the end of a road, but rather the beginning of one: a road on which figuring out solutions is the cornerstone of one's work. We encourage you to learn, know your facts, improve your mental computation skills, and ultimately, to use all this arsenal in solving problems from scratch. Once you start feeling the power of a good problem solver you will no longer think of a perfect score as a dream. You will be confident in your skills, you will continue to use your problem solving skills when you head towards college and graduate school, and you will end up cherishing the core of logical reasoning and the work stamina that makes you intrinsically more valuable than other applicants.

We believe in getting the most out of the effort we put in. We are not thrilled by the idea that a student might, with a little bit of luck, do above average on the SAT and *postpone* the real work of actually becoming academically valuable. That's because tomorrow comes inevitably, and we prefer to make sure we are ready for it. That means we must have skills. That's the bottom line. Plugging in all the answers is not a skill. Eliminating options is not a skill. The idea that the student can leverage the structure of the test itself to win points is fundamentally flawed: it leaves the student vulnerable to any changes in the presentation of the problems. The SAT is not cast in stone. The test evolves continuously and adapts to changes in the students' work philosophy. The only strategy that will never fail is to know the material well and to have sufficiently well developed problem solving skills so as to be able to actually solve the problems.

SAT problems are written with the goal of throwing the student off the beaten path. Instead of asking the student to apply a concept, a problem might ask the student to figure out whether the concept is applicable. The test puts you in charge, makes you responsible for the decision to apply steps you have seen before, to modify these steps to fit the new situation, or to figure out a new procedure altogether. This is why we believe that practice is the best strategy for preparation. You have to solve, solve, solve, until you feel comfortable making decisions, until you know your answers *must* be correct!

Our problems are designed to be close to SAT questions. In keeping with our philosophy, we design the problems to *anticipate* a number of format changes that are inevitable. Many format changes are not difficult to anticipate. For instance, calculators are capable of accomplishing more tasks than ever before. Only ten years ago, not every pocket calculator was able to understand parenthesis in expressions, produce answers in fraction form, or calculate logarithms with any base. All this has changed. With the advances in technology, many older SAT problems are immediately solved using a calculator. The exam changed too, and the simple use of symbols instead of numbers in an expression will require human work. It is a fallacy to assume that SAT questions are immutable. It is also a fallacy to rely entirely on previous SAT problems, the so-called "real SAT problems." They were real *yesterday*! Tomorrow is different and has different challenges. If a problem has been made public by the College Board, it is because it no longer has use for them. The currently released problems do not even form a large enough statistical sample to allow anyone to derive a complete strategy based on the "spirit of the SAT." The College Board is not dead - there are teams of very qualified people at their end who constantly adjust these problems to make sure only so many students get a certain score. Their work and analysis takes into account a variety of cultural shifts, such as the increased helpfulness of calculators, the "revelations" popularized by "quick fix" trainers, the lists of "SAT words," etc. The SAT does not have an obligation to use certain words. If it does so, it is based on their own statistics, and reflects their anticipation that only so many students will know or be able to infer the meaning of some word.

This workbook helps you practice by looking at the concepts you know from school in a different way: questioning their validity in the unfamiliar contexts, figuring out how to apply them to unusual situations, and building correct reasonings from scratch. These are the only skills that will *remain the same* as the exam evolves over time, the only skills that will *remain useful for other standardized exams*, and that will *not become obsolete during your lifetime*. Our questions are designed to ensure your readiness for *tomorrow's* SAT!

2.1   The SAT Preparation Workflow

This is a practice book, not a comprehensive guide. The market is replete with comprehensive guides both for pay and for free. This book does not teach you the concepts, nor does it list strategies for solving. It is a practice book with full solutions. It offers a variety of SAT-like tests, problems, as well as skill-building quizzes. We recommend this book as a supplementary material in addition to any class, prep book, or free SAT resource. Since "practice makes perfect," practicing with the problems in this workbook will enable you to ace your SATs!

2.2   Mastering the Concepts and Strategies for SAT Problem Solving

We recommend using a variety of resources. Free online resources such as www.mathinee.com, the Sparks notes, and the resources available at www.collegeboard.org are examples of sites where the concepts are listed from a variety of angles.

The mathematical facts and concepts used in SAT problems are not as simple as it may seem at first glance. The simplicity of the problem statements is often deceptive. The questions differ from textbook questions in a radical way.Textbook questions are designed to help you obtain a good grade and rely on repetitive, mechanical application of theory, direct connection between concept and its application. SAT questions are designed to help you fail.

SAT questions are like a battlefield where constant watchfulness, attention to detail, and rapid accurate responses are required for survival.

To do very well on the SATs, a student must first of all understand the nature of the exam. Familiar concepts will be used in unfamiliar situations, where one small detail can derail attempts at solving mechanically. Answer choices are carefully crafted to lure the student into a decision path that is incorrect. Common errors, such as applying analogies with other problems or errors resulting from an apparent similarity of form, will be featured as attractive answer choices.

Learn the concepts with an open mind. Do not focus on learning strategies based on steps! Instead, train yourself to solve each problem from scratch. SAT problems are constructed so as to expose and punish the mechanical application of step-based procedures. Try to make sure you have a good reason to apply an operation and you will do well on any reasoning test. While you train, solve problems slower and try to develop a complete understanding of the process of solving them.

2.3 What you Need: Accuracy, Speed, and Stamina

What many SAT candidates fail to realize is that the abilities needed to ace the exam are strongly interconnected. It is not sufficient to be able to solve all the problems correctly, one must also solve them rapidly and maintain that accuracy and rapidity for hours at a time!

Accuracy is the first goal of your training. The ability to comprehend the statement, observe and remember all the crucial details that may be different from routine, figure out a strategy, and implement that strategy correctly, is the foundation of your progress. It is by no means something that can be achieved by rushing through problem sets. To achieve a uniformly accurate response to SAT problems, you have to spend the first part of your training solving slowly, thinking through all the details very carefully, and experimenting with multiple strategies for each problem.

Writing down parts of your solution, making a diagram, a table, or a drawing should be part of the process of working on the solution. Writing helps you keep a record of your solving decisions and computations. If you suspect an error, this record is extremely useful in finding it - it will always be faster to find the error than to solve again from the top. Jotting down notes while solving is so useful, that you will find there is rapid improvement as soon as you start using some scratch paper. Even though we advise you to improve your mental computation skills, we totally disapprove of the students who expect to find a solution just by looking at the exam paper. Especially as you practice in order to improve your accuracy, taking notes and making drawings is one of the most useful work habits.

Speed is a different goal and should never be your first goal. The first goal is accuracy. Speed necessarily comes second and should only be a concern once good accuracy is already in place.

To achieve a perfect score, your have to achieve an average speed of one problem per minute. This includes: reading, comprehending, solving, and filling in the answer choice.

To achieve a competitive speed on exams such as the SAT, you have to:

- Try to find a shorter solution for each problem. The ideal solutions are those that can be implemented completely mentally. The less we use paper, pencil, and calculator, the faster we are. While this is not possible for all the problems, it is for many of them. Refine all your solutions from phase I training, to reflect this goal of extreme simplicity. Solving a problem in several ways and comparing the complexities of each step seems to be an incredible waste of time. But it is not, it is actually time that is incredibly well spent. The benefit of doing this will be so amazing and so solid that not for a moment should you doubt your decision.

- Eliminate guessing and plugging in answers as solving strategies. They are either risky or time consuming. There are many problems that I can solve a problem in my head in a few seconds - but if I start plugging answers in, I could be working for several minutes on each problem.

- Train yourself to recognize a lot of numbers that are significant in problem solving: perfect squares, square roots, prime numbers, Pythagorean triples, small factorials, and ratios of lengths in special triangles. Also make sure your mental computation ability is very good, as it is always faster to compute mentally than to enter data in the calculator!

- Reduce the use of the calculator to a minimum. Entering data is time consuming and prone to error. Any computation done with a calculator must be performed at least twice, to make sure no data entry errors have been made.

- Practice factoring quadratic expressions.

Stamina is extremely important on the SAT due to the duration of the test. You have to work at the same level of alertness for the entire duration of the exam. The goals you have to achieve in building the mental stamina for this test are:

- Understanding how much energy you need to get through the test.

- Training the capacity to focus on test questions for hours at a time.

- Acquiring the ability to expend the needed energy at a constant rate.

The SAT reasoning exam is like a marathon. Train for it as if you were training for a marathon and you will achieve impressive results that will empower you for a lifetime.

2.4 How to Use this Practice Book

This book is designed to give you full control over your progress.

Take each diagnostic test in the spirit of your current training phase. If you are training for accuracy, take your time solving and refining each solution. If you are training for speed, use a timer and try to solve in the suggested time.

Use the answer key to diagnose your response and figure out which training goals are the ones you must focus on next.

Take the short quizzes to build a good response time on individual skills.

Good luck on your SAT reasoning exam!

How to Interpret the Diagnostic Test

Take the diagnostic section in the allocated time and in conditions similar to the ones at the real test. Turn to the answer page and self-grade your test. Fill out the checkboxes on the diagnostic form. Since we work section by section we will use a simplified monitor of success which is the percentage of correct answers. You have to bring this to 100%!

Looking at the form, complete the diagnostic table with answers to the following questions:

- Are your incorrect answers mostly on questions in the first part of the test? YES/NO

- Are your incorrect answers mostly on questions that have difficulty levels 3 and 4? YES/NO

- Are your incorrect answers mostly from a specific topic (geometry, algebra, arithmetic)? YES/NO

- Are your incorrect answers somewhat evenly spread throughout the test? YES/NO

Now, look at the solutions for the questions you answered incorrectly, and answer the questions:

- Are your incorrect answers mostly due to minor errors? Would you have given a correct answer if there was only one question to answer and something important to you had been at stake?

- Are your incorrect answers mostly due to numeric errors? Were you able to figure out the question conceptually but had errors when executing the computations?

The answers to these questions can be interpreted to help you see *rapid progress* in your scores! Read on.

Are your incorrect answers mostly on questions in the first part of the test?

If yes, then it took you a while to become completely focused on the test. This was especially detrimental since the first questions are usually the easiest and there you were, rushing and getting them wrong.

To obtain a perfect score on this test, one must focus completely on the test and maintain that focus, while operating at top capacity for the whole duration of the test (which is, with administration and breaks, about 4 hours in total). When starting the exam, as well as after each administrative break, it is difficult to focus 100% as soon as the clock starts ticking. We have either been listening to music, talked to others, or started to snooze while the instructions were being read to us for the third time around. Like a vehicle, which needs a while to accelerate to freeway speed, your brain needs some time to focus and become completely immersed in the test.

How do you fix this?

- On the way to the test, do not listen to music, read a book, or watch a movie. Do not engage in a loud conversation with friends before entering the exam room.

- Put all your electronics away until the end of the test.

- Start thinking about the task at hand as soon as you leave your home. Stay calm and silent, removing the usual inputs of information from the next hour of your life.

- Visualize the first moments of the test. Visualize your level of focus and energy attaining a sustainable maximum as soon as the timer starts.

- Ignore all distractions during the test.

- Visualize yourself repeatedly starting each section of the test completely focused and ready to execute.

Are your incorrect answers mostly on questions that have difficulty levels 3 and 4?

This means you need more study and practice.

How do you fix this?

Get all the practice you can. Squeeze time from all possible corners and practice. If you have to live outside of your social networks for a few months, do so!

Are your incorrect answers mostly from a specific topic (geometry, algebra, arithmetic)?

This means you must target some specific areas of improvement.

How do you fix this?

Our free website www.mathinee.com is an excellent resource for short and to the point review of selected topics.

Are your incorrect answers somewhat evenly spread throughout the test?

This indicates that your ability to focus is uneven. After solving a question correctly your brain gets rapidly tired and 'relaxes' on the next question. The result is a wrong answer. You have to train yourself to maintain the same level of focus and heightened intellectual alertness for the whole time you are answering questions. Visualize yourself in a steady danger and practice through meditation.

This also shows you may be guessing on too many of the questions. In fact, you should not employ guessing *at all*. All questions have a perfectly rational answer. Solve them from the top, without looking at the answer choices. Once you have a solution, match it to the answer. This is the fastest, most reliable way to ace this test. Preparers who work 'by elimination' and 'by making intelligent guesses' are wasting your time. Literally. Because eliminating and deciding how to guess takes longer, in seconds, than working from the top. If you are good, that is. This book will help you become that good and ace this test with complete confidence.

How do you fix this?

- Visualize yourself focusing constantly and evenly.

- Visualize yourself in a prolonged situation in which you cannot afford to relax.

- Stop guessing, working by elimination, plugging answers back in, and making 'intelligent' guesses.

- Visualize yourself in control and under sustained fire. You cannot back out, you cannot relax. You must prevail.

Are your incorrect answers mostly due to minor errors?

We are familiar with this. From our experience as instructors, about 40% of students' errors fall in this category. We have run statistics that show this and we have obtained sudden increases in scores just by pointing this out to students.

By paying attention you can decrease the number of your errors by almost 50%! Isn't this incredible? You thought you needed to take a class, spend numerous hours taking practice tests, and your score could have improved substantially just by learning to focus better.

How do you fix this?

- Visualize yourself focusing.

- Every time you brush your teeth, remind yourself that you can improve your score substantially just by paying more attention.

- Once a day, put all electronics away and focus fully on a single thought. However simple, remain focused on that same thought for a short while.

If you do this, you will see your scores improving suddenly by a lot of points.

Remember though, that this is a one-time miracle improvement and that, to eliminate the remaining 60% of errors, greater mastery of the subject matter is needed.

DIAGNOSTIC ONE

This math section contains 18 questions to be solved in 25 minutes.

You are allowed to use a calculator. However, in order to minimize the time needed to complete the test, it is important to use the calculator as little as possible! Keep in mind:

- Entering operations in the calculator is more time consuming than performing the operations mentally.

- Data entry errors will be made in addition to other errors. Entering data is, in itself, a possible cause for error.

You can reduce the use of the calculator by memorizing well a short list of commonly used numbers: perfect squares from 1 to 20^2, powers of 2 from 2^0 to 2^{10}, some frequently used Pythagorean triples, etc. The list can be found in Appendix A, as well as on www.mathinee.com.

As you solve problems, build on the ability to recognize the numbers in this list. Every time you encounter them, tell yourself "hey, this is a power of 2", or "hey, this is a Pythagorean triple." After a short while, you will have memorized the list almost completely. But going over the list 'cold' a few times is also helpful, so go ahead and open Appendix A!

Before starting to solve, spend a few seconds to think about focusing, about making sure that you are ready to pay attention to every detail.

1. 5.61×10^{-5} is equal to which of the following:

(A) 561×10^{-3}

(B) $.561 \times 10^{-6}$

(C) 56.1×10^{-6}

(D) 56.1×10^{-4}

(E) $.561 \times 10^{6}$

Ⓐ Ⓑ Ⓒ Ⓓ Ⓔ

2. If a truck traveled for 5 hours at an average speed of 52 miles per hour, how fast must it travel for the next 3 hours in order to achieve an overall average speed of 55 miles per hour?

(A) 56

(B) 58

(C) 60

(D) 62

(E) 63

Ⓐ Ⓑ Ⓒ Ⓓ Ⓔ

3. On a farm, the ratio of cows to goats is $1 : 5$ and the ratio of cows to pigs is $2 : 3$. What is the ratio of pigs to goats?

(A) $3 : 5$

(B) $3 : 10$

(C) $15 : 2$

(D) $10 : 3$

(E) $6 : 5$

Ⓐ Ⓑ Ⓒ Ⓓ Ⓔ

4. A line with slope 5 is parallel to another line with y-intercept -1. The distance between their x-intercepts is 1. What is the y-intercept of the first line?

(A) -6 or 4

(B) 6 or -4

(C) 4 or -4

(D) 6 or -6

(E) 4

Ⓐ Ⓑ Ⓒ Ⓓ Ⓔ

5. Which of the answer choices is not a term of the sequence:

$$7, 11, 15, 19, \ldots$$

(A) 51

(B) 155

(C) 191

(D) 213

(E) 339

Ⓐ Ⓑ Ⓒ Ⓓ Ⓔ

6. The ratio of the width of a rectangle to its length is 4 : 11, and its perimeter is 75 feet. What is its area in square feet?

(A) 30 **(B)** 150 **(C)** 162.5

(D) 225 **(E)** 275

Ⓐ Ⓑ Ⓒ Ⓓ Ⓔ

7. The height of the waves at the Linda Mar Beach has been plotted against the temperature of the air.

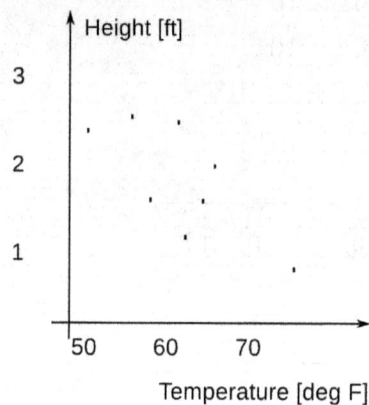

(A) high positive correlation

(B) high negative correlation

(C) low positive correlation

(D) low negative correlation

(E) no correlation

Ⓐ Ⓑ Ⓒ Ⓓ Ⓔ

8. The solutions of the inequality $-15x \geq 73$ are best described by which of the following inequalities?

(A) $x > -4.8$

(B) $x \geq -4.9$

(C) $x \leq -4.8$

(D) $x \leq -4.7$

(E) $x \leq -4.9$

9. A printer costs 200 dollars to purchase and .17 cents per page to operate. Another printer costs 260 dollars to purchase and .15 cents per page to operate. After printing how many pages will the extra cost for the second printer be amortized?

(A) 1600

(B) 2000

(C) 2600

(D) 3000

(E) 5200

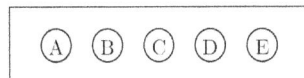

10. Each of the two bases of an isosceles trapezoid is tangent to a circle. The trapezoid has an area of 68 square units and the segment that connects the midpoints of the non-parallel sides has a length of 17 units. How many units long is the diameter of the circle?

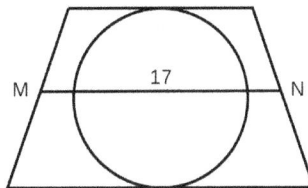

(A) 4

(B) 6

(C) 8

(D) 8.5

(E) 34

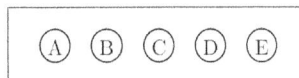

11. 21% of the volume of a container is taken by tennis balls and the rest is air. If we remove 10 tennis balls, air represents 85% of the volume of the container. How many tennis balls were in the container to start with?

(A) 31

(B) 32

(C) 33

(D) 34

(E) 35

Ⓐ Ⓑ Ⓒ Ⓓ Ⓔ

12. Two vertices of a square have coordinates $(5, 4)$ and $(5, -1)$. Using the same unit of length, the area of the figure formed by the square and its reflection across the x-axis is:

(A) 30

(B) 40

(C) 45

(D) 50

(E) 100

Ⓐ Ⓑ Ⓒ Ⓓ Ⓔ

13. On Wall Street there live 54 families. 24 families own a Chevrolet and 39 families own a Ford. How many families own at least 2 vehicles?

(A) 9

(B) 14

(C) 15

(D) 25

(E) 30

Ⓐ Ⓑ Ⓒ Ⓓ Ⓔ

14. When defending against a penalty shot, a soccer goalie has an average rate of success of 60% for shots that arrive in the area that is shaded in the figure, and a rate of success of 30% for shots that arrive in the area that is left white. If all the shots are aimed at the goal and the probability of arriving at any one point of the defended area is the same, which answer is closest to the probability of scoring a penalty shot? (The figure is not to scale and lengths are measured in feet.)

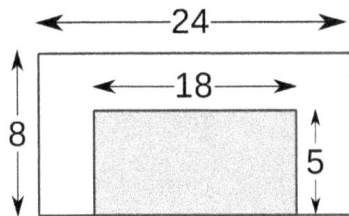

(A) 45%

(B) 50%

(C) 56%

(D) 60%

(E) 90%

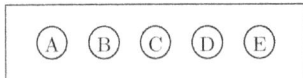

15. A function $f(k)$ is defined for any positive integer. If it satisfies the conditions:

$$f(1) = 10$$
$$f(3k) = f(k) + 3$$

what is the value of $f(3^{10})$?

(A) 37

(B) 40

(C) $90 \cdot 3^5$

(D) $3^{10} - 10$

(E) $10 + 3^{10}$

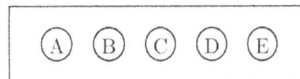

16. Two identical circles are transformed. The radius of one of them increases by 15% and the radius of the other one decreases by 5%. By approximately how much percent is the area of the larger circle greater than the area of the smaller circle?

(A) 10

(B) 20

(C) 42

(D) 47

(E) 75

(A) (B) (C) (D) (E)

17. A point in the plane has rectangular coordinates x and y that satisfy $x - y > 4$. This point can be only:

(A) in quadrant IV

(B) in quadrants I and IV

(C) in quadrants I, III and IV

(D) in quadrants I, II, III, and IV

(E) in quadrants I and III

(A) (B) (C) (D) (E)

18. The two x-intercepts of the parabola $x^2 - 2x - 15$ are also points on a circle. The center of the circle has coordinates:

(A) (any, any)

(B) (any, 1)

(C) (1, any)

(D) (1, 0)

(E) (4, 0)

(A) (B) (C) (D) (E)

19. In a survey about their favorite type of vacation, people's responses were summarized in a circle graph:

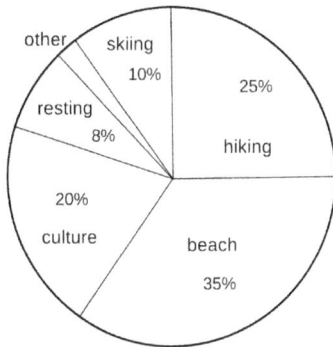

If the number of people surveyed who chose hiking is larger by 510 than the number of people who chose to rest, how many people chose skiing as their favorite vacation activity?

(A) 170

(B) 300

(C) 430

(D) 590

(E) 3000

Ⓐ Ⓑ Ⓒ Ⓓ Ⓔ

20. A data set consisting of integer values has a mode of 7 and a median of 15. What is the smallest possible range of the data?

(A) 7

(B) 8

(C) 9

(D) 10

(E) 22

Ⓐ Ⓑ Ⓒ Ⓓ Ⓔ

4.1 Self-grade Diagnostic One

Question	1	2	3	4	5	6	7	8	9	10	11	12
Level	1	1	1	2	1	2	1	2	2	3	3	3
Topic	N	N	N	A	N	G	S	A	N	G	N	G
Answer	C	C	B	A	D	E	D	E	D	A	E	B
Correct												
Incorrect												
Skipped												

Question	13	14	15	16	17	18	19	20
Level	3	4	3	3	3	3	4	4
Topic	S	S	A	G	A	A	S	S
Answer	A	C	B	D	C	C	B	C
Correct								
Incorrect								
Skipped								

Are your incorrect answers:

mostly on questions in the first part of the test?	Yes	No
mostly on questions that have difficulty levels 3 and 4?	Yes	No
mostly from a specific topic (geometry, algebra, arithmetic)?	Yes	No
somewhat evenly spread throughout the test?	Yes	No
mostly due to minor errors?	Yes	No
mostly due to numeric errors?	Yes	No

A-algebra, N-number sense, G-geometry, S-statistics

4.2   Diagnostic One Solutions

Question 1 To obtain an equivalent decimal we have to multiply and divide at the same time by the same factor. This is true only about choice (C):

$$5.61 \times 10^{-5} = 5.61 \times 10 \times 10^{-1} \times 10^{-5}$$
$$= 56.1 \times 10^{-5}$$

Question 2

The average speed is the ratio of the total distance and the total time traveled. The truck traveled $52 \cdot 5 = 260$ miles in the first 5 hours. The total distance it must travel that day is $55 \cdot 8 = 440$ miles. The remaining $440 - 260 = 180$ miles must be covered in 3 hours, at an average speed of $180 \div 3 = 60$ miles per hour.

Question 3

Pigs to cows is $3 : 2$ and cows to goats is $2 : 10$. Therefore, the ratio of pigs to goats is $3 : 10$.

$$\frac{P}{C} \times \frac{C}{G} = \frac{P}{G} = \frac{3}{2} \times \frac{1}{5} = \frac{3}{10}$$

Question 4

The second line has equation:

$$y = 5x - 1$$

The first line has equation:

$$y = 5x + b$$

where b is not known. The x-intercepts of the two lines are, respectively:

$$x_1 = \frac{1}{5}$$

$$x_2 = -\frac{b}{5}$$

The distance between the two x-intercepts is the positive difference between the intercepts $|x_1 - x_2|$:

$$x_1 - x_2 = 1$$

$$\frac{1}{5} + \frac{b}{5} = 1$$

$$1 + b = 5$$

$$b = 4$$

or

$$x_2 - x_1 = 1$$

$$-\frac{b}{5} - \frac{1}{5} = 1$$

$$-b - 1 = 5$$

$$b = -6$$

The y-intercept of the first line is either -6 or 4.

Question 5

The sequence is an arithmetic sequence with first term 7 and common difference 4. Subtract 7 from each choice. If the result is not a multiple of 4 we are not dealing with a term of the sequence.

$$51 - 7 = 44$$

$$155 - 7 = 148$$

$$191 - 7 = 184$$

$$213 - 7 = 206$$

$$339 - 7 = 332$$

Of these, we could actually have stopped at choice (D), since 206 is clearly not divisible by 4.

An integer is divisible by 4 if its last two digits form a number divisible by 4.

Question 6

If the width is $4x$ and the length is $11x$, then the perimeter is equal to:

$$2 \cdot 4x + 2 \cdot 11x = 2 \cdot 15x = 30x$$

Since we also know that the perimeter is 75 feet long, we can find x:

$$\begin{aligned} 30x &= 75 \\ 2x &= 5 \\ x &= 2.5 \end{aligned}$$

The width and length are, in feet:

$$\begin{aligned} W &= 4x = 4 \cdot 2.5 = 10 \\ L &= 11x = 11 \cdot 2.5 = 27.5 \end{aligned}$$

The area, in square feet, is:

$$W \cdot L = 27.5 \cdot 10 = 275$$

Question 7

The data has low negative correlation.

Question 8

The inequality has the solution:

$$\begin{aligned} -15x &\geq 73 \\ 15x &\leq -73 \\ x &\leq -\frac{73}{15} \end{aligned}$$

Note that, changing the signs in the inequality has reversed the inequality sign.

Using the calculator to compute the value of the fraction we obtain $4.8\overline{6}$. Figure on the number line the position of this number compared to the positions of the numbers provided as answer choices:

$$-4.9 < -4.8\overline{6} < -4.8 < -4.7$$

All the numbers smaller than or equal to -4.9 are solutions of the inequality. All the other choices include only some or none of the solutions. Among the given choices, the answer choice (E) is the best description of the solution.

Question 9

Denote the number of printed pages by x. After printing x pages, the total cost of owning and operating the printers is, respectively:

$$.17x + 200 \qquad \text{for the first printer}$$

$$.15x + 260 \qquad \text{for the second printer}$$

After the two costs become equal, the second printer will be cheaper to operate henceforth:

$$\begin{aligned} .17x + 200 &= .15x + 260 \\ .02x &= 60 \\ 2x &= 6000 \\ x &= 3000 \end{aligned}$$

After 3000 pages, the second printer print at a lower cost per page than the first printer.

Question 10

The segment that connects the midpoints of the two non-parallel sides of a trapezoid has the length equal to the average of the two bases.

The area of the trapezoid is equal to the height of the trapezoid multiplied by the average of the two bases.

Divide the area of the trapezoid by the average of the two bases to obtain the height:

$$68 \div 7 = 4$$

This is also the diameter of the circle.

Question 11

Denote the number of tennis balls by N and the volume of the container by V. The percent volume occupied by a single ball is:

$$\frac{21}{100} \cdot V \div N = \frac{21V}{100N}$$

After removing 10 balls, the percent volume occupied by a single ball is:

$$\frac{15V}{100(N - 10)}$$

The two must be equal:

$$\frac{21V}{100N} = \frac{15V}{100(N - 10)}$$

$$21\cancel{V} \times \cancel{100}(N - 10) = 15\cancel{V} \times \cancel{100}N$$

$$21(N - 10) = 15N$$

$$21N - 15N = 210$$

$$6N = 210$$

$$N = 35$$

Question 12

The square and its reflection form a rectangle with dimensions 5 and 8. Its area is 40 square units. A possible figure is:

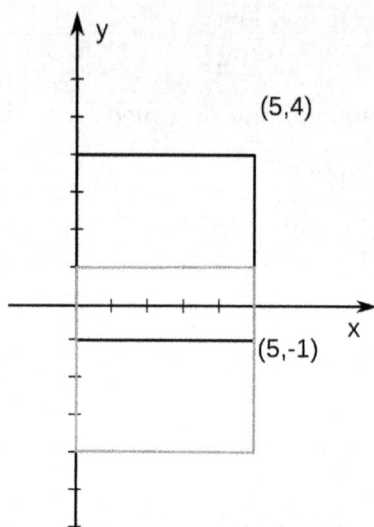

Note that the square can be drawn on the right side of the line $x = 5$ but the answer remains unchanged.

Question 13

Make a Venn diagram and apply the inclusion-exclusion principle:

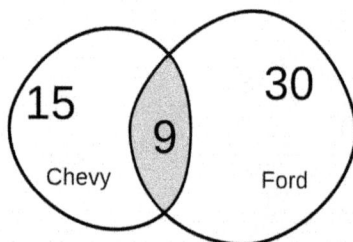

Since there are $24 + 39 = 63$ vehicles of the two brands, if there are no other vehicles, then there are $63 - 54 = 9$ families who own both brands.

Question 14

Denote the desired probability with x. The total area of the goal is $8 \times 24 = 192$ square feet. The grey area is $18 \times 5 = 90$ square feet. The white area is $192 - 90 = 102$ square feet. Of 192 shots, 90 will arrive in the grey area. Of these $90 \times 0.6 = 54$ are intercepted and $90 - 54 = 36$ score a goal. Of 192 shots, 102 will arrive in the white area and $102 \times 0.3 = 30.6$ are intercepted and $102 - 30.6 = 71.4$ score a goal. In total, of 192 penalty shots, $71.4 + 36 = 107.4$ score a goal. The probability of scoring is:

$$\frac{107.4}{192} \approx 55.83\%$$

Question 15

Each time we multiply the argument by 3, the value of the function increases by 3:

$$
\begin{aligned}
f(1) &= 10 \\
f(3) &= 10 + 3 \\
f(9) &= 10 + 3 + 3 \\
\cdots &= \cdots \\
f(3^{10}) &= 10 + 3 \times 10 = 10 + 30 = 40
\end{aligned}
$$

Question 16

Let the identical circles have radius 1. After 15% increase, the radius of one circle becomes 1.15. After a 5% decrease, the radius of the other circle becomes 0.95. The difference between the larger area and the smaller area is:

$$\pi \cdot 1.15 \cdot 1.15 - \pi \cdot 0.95 \cdot 0.95$$

The percent difference is:

$$
\begin{aligned}
\frac{\pi \cdot 1.15 \cdot 1.15 - \pi \cdot 0.95 \cdot 0.95}{\pi \cdot 0.95 \cdot 0.95} &= \frac{(1.3225 - 0.9025)\pi}{0.9025\pi} \\
&= \frac{0.42}{0.9025} \approx .4653
\end{aligned}
$$

The area of the larger circle is approximately 47% larger.

Question 17

The inequality can be written as:

$$x - 4 > y$$

$$y < x - 4$$

and the solution set is the half-plane below the line $y = x - 4$:

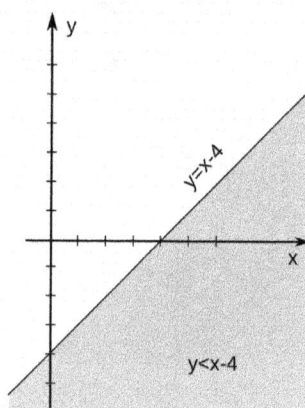

Question 18

The axis of symmetry of the parabola is the perpendicular bisector of the segment that separates the two x-intercepts of the parabola: -3 and 5. The center of a circle that has two points in common with a segment is on the perpendicular bisector of that segment. It follows that the center of the circle is anywhere on the axis of symmetry of the parabola. It has the same x-coordinate as the vertex of the parabola:

$$x_v = -\frac{-2}{2} = 1$$

The y-coordinate of the center of the circle can be anywhere on the vertical line $x = 1$.

Question 19

Assume the total number of people surveyed was N. Then $\dfrac{25}{100}N$ is the number of people who chose hiking and $\dfrac{8}{100}N$ is the number of people who chose resting. The difference must equal 510:

$$\frac{25}{100}N - \frac{8}{100}N = 510$$

$$\frac{17}{100}N = 510$$

$$\frac{17}{100}N = 3 \cdot 17 \cdot 10$$

$$N = 30 \cdot 100 = 3000$$

3000 people answered the survey. 300 of them chose skiing as their favorite vacation activity.

Question 20

Note that the number of data points is not specified. If we increase the number of data points to 7, we can form the set:

$$7, 7, 7, 15, 15, 16, 16$$

which has a range of only 9. The *range* is the difference between the largest value and the smallest value in the set.

ALGEBRA QUIZ ONE

1. Fill in:

$$(a + b)^2 = \cdots$$
$$(a - b)^2 = \cdots$$

2. Match the items:

(a) $(1 - x)^2$ **(x)** $7^2 - 14 \cdot 9 + 9^2$

(b) $(5 + 6)^2$ **(y)** $2hp + p^2 + h^2$

(c) $(7 - 9)^2$ **(z)** $1000^2 + 4004$

(d) $121x^2 - 44xy + 4y^2$ **(w)** $25 + 36 + 60$

(e) 1002^2 **(v)** $x^2 - 2x + 1$

(f) $(p + h)^2$ **(t)** $(2y - 11x)^2$

3. Answer rapidly, using memorized facts:

(a) $13^2 =$ **(b)** $11^3 =$

(c) $2^7 =$ **(d)** $3^3 =$

(e) $3^2 + 4^2 =$ **(f)** $2^{10} =$

(g) $5^3 =$ **(h)** $25^2 =$

(j) $15^2 =$ **(k)** $12^2 + 5^2 =$

Answers to questions 1-3

1. The squares of the sum and difference binomials are:

$$(a + b)^2 = a^2 + 2ab + b^2$$
$$(a - b)^2 = a^2 - 2ab + b^2$$

2. The matches are:

$$(a) \rightarrow (v)$$
$$(b) \rightarrow (w)$$
$$(c) \rightarrow (x)$$
$$(d) \rightarrow (t)$$
$$(e) \rightarrow (z)$$
$$(f) \rightarrow (y)$$

3. Refer to Appendix A for facts that should be memorized for easy recognition:

(a) $13^2 = 169$ **(b)** $11^3 = 1331$

(c) $2^7 = 128$ **(d)** $3^3 = 27$

(e) $3^2 + 4^2 = 5^2$ **(f)** $2^{10} = 1024$

(g) $5^3 = 125$ **(h)** $25^2 = 625$

(j) $15^2 = 225$ **(k)** $12^2 + 5^2 = 13^2$

GEOMETRY QUIZ ONE

1. The midline of a triangle is ...... to a side of the triangle. Its length is ...... of the length of the side it is ...... to.

2. Any point on the perpendicular bisector of a segment is ...... from the two ends of the segment. Such a point forms an ...... triangle with the segment as a base.

3. If an angle inscribed in a circle intercepts the diameter, its measure is equal to ...... degrees of arc.

4. The line tangent to a circle is ...... to the radius that connects the center to the point of tangency.

5. The length of a diagonal of a square is ...... times the length of the side.

6. If two circles are tangent interior, their centers and the point of tangency are ......

7. From a point exterior to a circle, but in the same plane as the circle, one can draw ...... lines tangent to the circle. The common tangent theorem states that the segments from the point to the respective points of tangency have the same ......

8. Any point on the angle bisector of an angle is ...... to the sides of the angle.

9. Two segments that are parallel and equal in length form a ......

10. The two diagonals of a parallelogram ...... each other.

Answers to questions 1-10

1. The midline of a triangle is *parallel* to a side of the triangle. Its length is *half* of the length of the side it is *parallel* to.

2. Any point on the perpendicular bisector of a segment is *equidistant* from the two ends of the segment. Such a point forms an *isosceles* triangle with the segment as a base.

3. If an angle inscribed in a circle intercepts the diameter, its measure is equal to 90 degrees of arc.

4. The line tangent to a circle is *perpendicular* to the radius that connects the center to the point of tangency.

5. The length of a diagonal of a square is $\sqrt{2}$ times the length of the side.

6. If two circles are tangent interior, their centers and the point of tangency are *collinear (on the same line)*.

7. From a point exterior to a circle, but in the same plane as the circle, one can draw *two* lines tangent to the circle. The common tangent theorem states that the segments from the point to the respective points of tangency have the same *length*.

8. Any point on the angle bisector of an angle is *equidistant* to the sides of the angle.

9. Two segments that are parallel and equal in length form a *parallelogram*.

10. The two diagonals of a parallelogram *bisect* each other.

Data Analysis Quiz One

1. How many samples have been collected to produce the histogram in the figure?

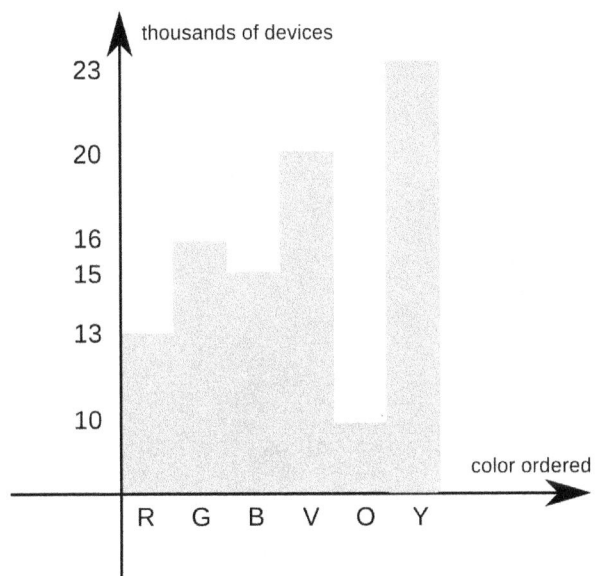

2. What is the range of each of the following data sets?

(a) $1.5, 9.2, 3, 7, 1, 10.7$

(b) $5, 5.5, 5.15, 5.515, 5.551, 5.1115$

(c) $10^3, 10^{\frac{1}{6}}, 10^0, 10^{-100}, 0$

(d) $0.001, 0.01, 0.09, 0.009, 0.091, 0.0019$

Answers to questions 1-2

1. Count the samples for each color and add. Make sure you notice the units used; in this case, thousands of devices.

$$(13 + 16 + 15 + 20 + 23 + 10) \times 1000 = 97,000$$

2. The range of a data set is the difference between the largest value and the smallest value.

(a) $10.7 - 1 = 9.7$

(b) $5.551 - 5 = 0.551$

(c) $10^3 - 0 = 1000$

(d) $0.09 - 0.001 = 0.089$

To identify rapidly the smallest and the largest values, the following facts are useful:

- A negative exponent does not make the number negative. It is not necessary to calculate or estimate 10^{-100}, since it is positive and therefore larger than 0, which is also a data point in the set.

- Any base raised to an exponent of zero equals 1.

- It is not necessary to calculate or estimate $10^{\frac{1}{6}}$, since it is positive. The fractional exponent indicates a sixth order root.

NUMBER SENSE QUIZ ONE

Statement	True	False	Sometimes true
Zero is even.			
Numbers divisible by 7 have a digit sum that is divisible by 7.			
Zero is negative.			
By subtracting a power of 2 from a power of 2 we get a power of 2.			
The result of division by zero is undefined.			
Zero is divisible by 7.			
-8 is an even number.			
There are 5 odd perfect squares from 1 to 100.			
1 is a prime number.			
By subtracting two perfect squares we obtain a perfect square.			
9, 12, 15 is a Pythagorean triple.			

Explanation	True	False	Sometimes true
Zero is even since $0 = 2 \times 0$	V		
Not generally true, but $7 = 1 \times 7$ and $0 = 0 \times 7$ are exceptions.			V
Zero is neither negative nor positive.	V		
Not generally true. Example: $2^1 - 2^0 = 2^0$. Counterexample: $2^3 - 2^0 = 7$.			V
The result of division by zero is undefined.	V		
Zero is divisible by 7 since $0 = 0 \times 7$.	V		
-8 is even since $-8 = 2 \times (-4)$.	V		
There are 5 odd perfect squares from 1 to 100: $1, 9, 25, 49, 81$.	V		
1 is not prime since it has only one divisor. 1 is an *improper prime*.		V	
Some numbers, such as Pythagorean triples behave in this way: $25 - 16 = 9$.			V
$9, 12, 15$ is a Pythagorean triple since is is a multiple of a Pythagorean triple $3 \times 3, 4 \times 3, 5 \times 3$. Check that $9^2 + 12^2 = 15^2$.	V		

Diagnostic Two

This math section contains 18 questions to be solved in 25 minutes.

You are allowed to use a calculator.

The first 8 questions are multiple choice, the remaining 10 questions are student response questions. For the student response questions, the answer consists of 4 characters chosen from the symbols: . /, and any of the ten digits. The answer may not start with a zero. Answers that have both a fractional and a decimal representation may be represented either way. For example, the representations:

$$1/4$$

and

$$.25$$

are equivalent. Note that 0.25 is not a possible entry.

Before starting to solve, spend a few seconds to think about focusing, about making sure that you are ready to pay attention to every detail.

1. If $x = -5$ and $y = -4$ what is the value of:

$$(x - 5)(7 + y)$$

(A) -30

(B) 0

(C) 20

(D) 30

(E) 110

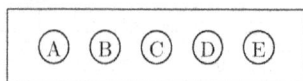

2. A suburban public transportation vehicle stops at every station for 1 minute. If the distances, in miles, between the stations increase like the sequence: $3, 5, 7, \ldots$, and the vehicle's speed between stops is 48 mph on average, approximately how many minutes did it take from the first departure to the fourth arrival?

(A) 4

(B) 33

(C) 34

(D) 40

(E) 41

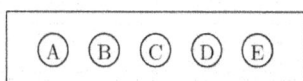

3. In the figure below all segments intersect at right angles. What is the perimeter of the figure?

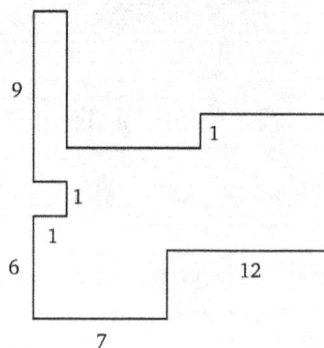

(A) 70

(B) 71

(C) 72

(D) 74

(E) cannot be determined

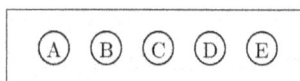

4. In the following data set, if the median value is 31 what is the mean?

$$\{13, 13, 28, 28, x, x, 58, 58\}$$

(A) 32.5

(B) 33.25

(C) 34.75

(D) 36

(E) 39

5. How many integer values does the function $f(x) = \dfrac{2}{7}x$ have if $-16 < x < 15$?

(A) 0

(B) 2

(C) 3

(D) 4

(E) 5

6. Which of the answers below is equivalent to $|x - y| + |y - x|$?

(A) 0

(B) $2x$

(C) $2y$

(D) $2x - 2y$

(E) $|2y - 2x|$

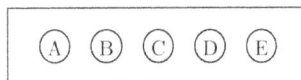

7. Erica's phone bills have increased by 10% in November, by another 20% in December but her bill for January was 8% lower than the December one. Which of the following is closest to her average percent increase over the three months in question?

(A) 7%

(B) 18%

(C) 20%

(D) 21%

(E) 22%

8. In the graph below the x-axis represents time and the y-axis the velocity, in feet per second, of a ball that bounces off a wall in the same direction it came from. How far from its starting point did the ball get during the 5 seconds depicted by the graph?

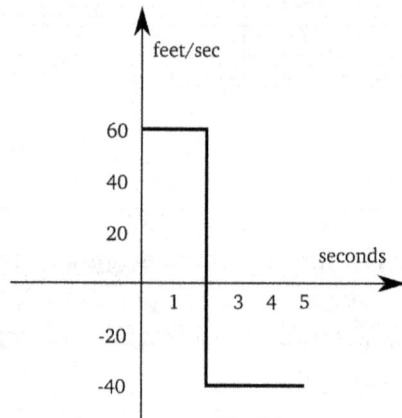

feet/sec

60

40

20

seconds

1   3 4 5

-20

-40

(A) 0

(B) 80

(C) 100

(D) 240

(E) 320

Ⓐ Ⓑ Ⓒ Ⓓ Ⓔ

9.   If

$$a = 3 + \frac{1}{3} + \frac{1}{9} + \frac{1}{27}$$

and

$$b = 3a - 1 - \frac{1}{3} - \frac{1}{9}$$

then b is:

10. $\frac{1}{3}t + t$ is larger than $\frac{1}{3}m + m$ by 16. How much larger than $\frac{1}{3}(t - m)$ is $t - m$?

11. Sandy forgot the last two digits of her password, but she remembers the password was a multiple of 25. What is the maximum number of passwords she must try until she stumbles upon the forgotten password?

12. A small population has filled out a survey about their daily intake of water. A histogram summarizes the data that was collected: how many people drink 4 glasses a day, 5 glasses a day, etc. On average how many glasses a day does this population drink? (Round to the nearest tenth.)

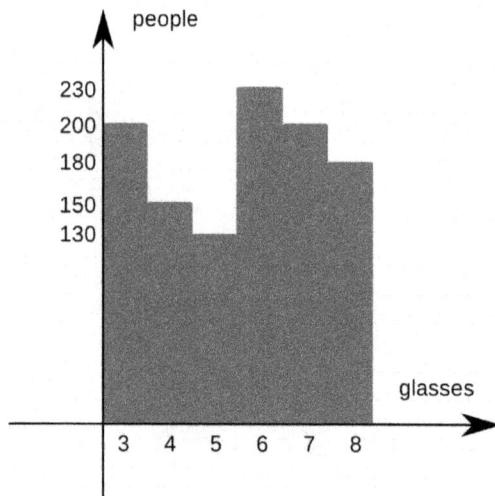

13.   A magician has a hat with sixty chips in it. Each chip is marked with a different number. A member of the audience is asked to remove a chip while blindfolded. His number is written down and the chip is not returned to the hat. Another member of the audience is asked to do the same. What is the probability that the number of the second person is smaller than the number of the first person?

14.   In a 8×8 table, we write in each cell the product of the line number and the column number of that cell. How many of the numbers in the table are perfect squares?

15.   The triangle ABC has area 18 square units. The point M is the midpoint of AB and the point N is such that $AN = 2NC$. Point P is the midpoint of BC and point Q is such that $BQ = 2QC$. How many square units is the shaded area? (The figure is not drawn to scale.)

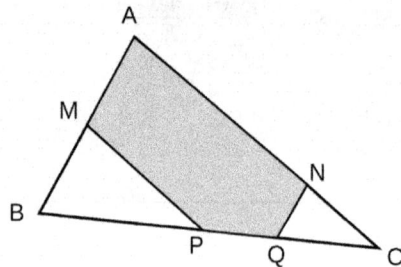

16.   A truck loaded a third of the merchandise. A second truck loaded half as much as the first truck. A third truck loaded twice as much as the second truck. A fourth truck loaded the remaining merchandise. What fraction of the load of the first truck was the load of the fourth truck?

17. If $(m+n)^2 = 5$ and $m^2 + n^2 = 5$, what is the value of $m \cdot n$?

18. The volume of a cone is 27π. If the diameter of its base circle is equal to twice its height, what would be the area of a parallelogram with the same height and base equal to the diameter?

9.1 Self-grade Diagnostic Two

Question	1	2	3	4	5	6	7	8
Level	1	1	1	1	2	2	3	3
Topic	A	N	G	S	A	A	N	A
Answer	A	B	D	B	E	E	D	A
Correct								
Incorrect								
Skipped								

Question	9	10	11	12	13	14	15	16	17	18
Level	2	2	2	3	4	3	4	4	3	4
Topic	N	A	N	S	S	N	G	N	A	G
Answer	9	8	3	5.6	0.5	10	11.5	0.5	0	18
Correct										
Incorrect										
Skipped										

Are your incorrect answers:

mostly on questions in the first part of the test?	Yes	No
mostly on questions that have difficulty levels 3 and 4?	Yes	No
mostly from a specific topic (geometry, algebra, arithmetic)?	Yes	No
somewhat evenly spread throughout the test?	Yes	No
mostly due to minor errors?	Yes	No
mostly due to numeric errors?	Yes	No

A-algebra, N-number sense, G-geometry, S-statistics

9.2   Diagnostic Two Solutions

Question 1

Substitute the values and follow the order of operations:

$$(-5 - 5)(7 - 4) = (-10) \cdot (3)$$
$$= -30$$

Question 2

The total distance from the first departure to the fourth arrival is:

$$3 + 5 + 7 + 9 = 24$$

The time it took the vehicle to travel 24 miles at an average speed of 48 mph is:

$$t = \frac{\text{Distance}}{\text{Speed}} = \frac{24}{48} = \frac{1}{2} \text{ hours} = 30 \text{ min}$$

The travel time was 30 minutes and there were three stops, each 1 minute long. The total time needed was 33 minutes.

Question 3

The perimeter is the same as that of a rectangle with sides 16 and 19 to which we must add the two indentations.

The bolded segments form two groups of vertical lengths, each $9 + 1 + 6 = 16$ units long:

The bolded segments form two groups of horizontal lengths, each $7 + 12 = 19$ units long:

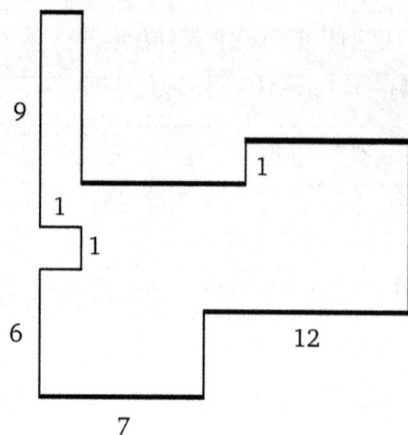

and the two indentations bolded below must be added on:

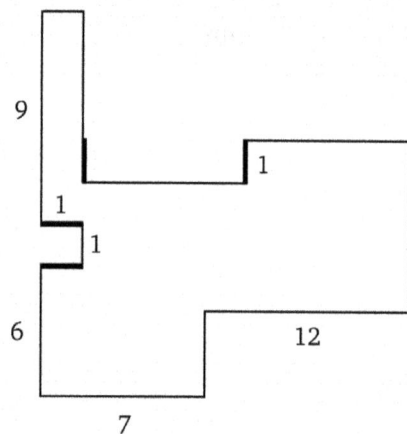

The perimeter is:

$$16 + 16 + 2 + 19 + 19 + 2 = 74$$

Question 4

There are 8 points in the data set. Therefore, the median is the average of the two middle points:

$$31 = \frac{x + 28}{2}$$

Solve for x to find $x = 34$. The mean is:

$$\frac{13 + 13 + 28 + 28 + 34 + 34 + 58 + 58}{8} = \frac{13 + 28 + 34 + 58}{4} = \frac{133}{4} = 33.25$$

Question 5

To obtain integer values for $f(x)$, x must be a multiple of 7. There are 5 multiples of 7 in the interval $-16 < x < 15$:

$$-14, -7, 0, 7, 14$$

Question 6

The absolute values are equivalent to:

$$|x - y| = x - y \text{ if } x \geq y$$
$$|x - y| = y - x \text{ if } x < y$$
$$|y - x| = y - x \text{ if } y \geq x$$
$$|y - x| = x - y \text{ if } y < x$$

Assuming $x \geq y$:

$$|x - y| + |y - x| = x - y + x - y = 2x - 2y = 2(x - y)$$

Assuming $x < y$:

$$|x - y| + |y - x| = y - x + y - x = 2y - 2x = 2(y - x)$$

Question 7

The percent change is 21.44%. Denote Erika's phone bill before November by B. The three variations in the billed amount will have the following effect on the bill:

$$B \cdot 1.1 \cdot 1.2 \cdot 0.92 = B \cdot 1.2144$$

which represents an increase of 21.44% of the billed amount.

Question 8

The ball traveled 60×2 feet towards the wall and 3×40 feet from the wall back. Overall, the ball just returned to its starting point.

Question 9

Notice that $3a$ will have some of the terms of b only with the opposite sign:

$$3a = 9 + 1 + \frac{1}{3} + \frac{1}{9}$$

Therefore, b is, at a glance, equal to 9.

Question 10

$$\frac{1}{3}t + t = (\frac{1}{3} + 1)t = \frac{4}{3}t$$

$$\frac{1}{3}m + m = \frac{4}{3}m$$

$$\frac{4}{3}t = \frac{4}{3}m + 16$$

$$\frac{4}{3}t - \frac{4}{3}m = 16$$

$$t - m = \frac{3}{4} \cdot 16$$

$$t - m = 12$$

$$\frac{1}{3}(t - m) = \frac{1}{3} \cdot 12 = 4$$

Therefore, $t - m$ is larger than $\frac{1}{3}(t - m)$ by 8.

Question 11

A number is divisible by 25 if its last two digits form a number divisible by 25. There are 3 2-digit combinations that are divisible by 25: 00, 25, and 75. Sandy must try three passwords.

Question 12

The average is given by the total number of glasses drunk divided by the total number of people:

$$\text{total glasses} = 200 \cdot 3 + 150 \cdot 4 + 130 \cdot 5 + 230 \cdot 6 + 200 \cdot 7 + 180 \cdot 8$$

$$\text{total people} = 200 + 150 + 130 + 230 + 200 + 180$$

Divide to obtain approximately 5.6.

Question 13

By taking two different numbers out of the hat without replacement, one of the numbers has to be smaller than the other. Therefore the probability for the first number to be larger is equal to the probability for the first number to be smaller than the second number. Both probabilities must add to 1 since it is certain that the number is either smaller or larger. Each probability must be 0.5.

Question 14

There are 8 perfect squares in cells that have the same row and column. But also, there are 2 additional perfect squares in cells where a number is multiplied by its cube: $2 \times 8 = 8 \times 2 = 16$. The total number of perfect squares is 10.

Question 15

Triangle $\triangle MBP$ is similar to $\triangle ABC$. Its side lengths are one half of the corresponding side lengths of $\triangle ABC$. Triangle $\triangle NQC$ is similar to $\triangle ABC$. Its side lengths are one third of the corresponding side lengths of $\triangle ABC$. Therefore, the area of $\triangle MBP$ is equal to one quarter of the area of $\triangle ABC$, and the area of $\triangle NQC$ is equal to one ninth of the area of $\triangle ABC$. The shaded area is:

$$\begin{aligned} A_{ANQPM} &= A_{\triangle ABC} - A_{\triangle ABC} \cdot \frac{1}{4} - A_{\triangle ABC} \cdot \frac{1}{9} \\ &= A_{\triangle ABC} \times \left(1 - \frac{1}{4} - \frac{1}{9}\right) = A_{\triangle ABC} \times \frac{36 - 9 - 4}{36} \\ &= A_{\triangle ABC} \times \frac{23}{36} \\ &= 18 \times \frac{23}{36} = 11.5 \end{aligned}$$

Question 16

Denote the merchandise with M (or assume it is a number, such as 1). The three trucks load:

$$\frac{1}{3} + \frac{1}{6} + \frac{1}{3} = \frac{2}{3} + \frac{1}{6} = \frac{5}{6}$$

The fourth truck loads $\frac{1}{6}$ of the total load. This is:

$$\frac{1}{6} = \frac{1}{2} \cdot \frac{1}{3}$$

one half of the load of the first truck.

Question 17

Use the algebraic identity:

$$(m + n)^2 = m^2 + n^2 + 2mn$$

It follows that $mn = 0$.

Question 18

The volume of the cone is:

$$\begin{aligned} \pi R^2 h &= 27\pi \\ R^2 h &= 27 \end{aligned}$$

The diameter is equal to twice the height h:

$$2R = 2h \quad \rightarrow R = h$$

Therefore:

$$R^2h \;=\; R^2 \cdot R = R^3$$

$$R^3 \;=\; 27$$

$$R \;=\; h = 3$$

The area of a parallelogram with the same height and base equal to the diameter is:

$$B \times h = 2h \times h = 2h^2 = 18$$

ALGEBRA QUIZ TWO

1. Match the items:

(a) 3^{3^4} (i) 3^9

(b) $(3^3)^4$ (j) 3^8

(c) 3^{2^3} (k) 3^{4^3}

(d) $(3^3)^3$ (l) 3^5

(e) $3^2 \cdot 3^3$ (m) 3^{81}

(f) 3^{64} (n) 3^{12}

2. Match the items:

(a) $125^{\frac{1}{3}}$ (i) 27

(b) $27^{\frac{2}{3}}$ (j) 2

(c) $121^{\frac{1}{2}}$ (k) 32

(d) $64^{\frac{1}{6}}$ (l) 9

(e) $81^{\frac{3}{4}}$ (m) 5

(f) $16^{\frac{5}{4}}$ (n) 11

3. Match the items:

(a) $9^{-\frac{1}{2}}$ (i) 16

(b) 2^{-3} (j) $\left(\dfrac{2}{3}\right)^{(-1)^{-1}}$

(c) $\left(\dfrac{1}{4}\right)^{-2}$ (k) $\dfrac{1}{8}$

(d) $\left(\dfrac{2}{3}\right)^{-1}$ (l) $\dfrac{1}{2}$

(e) $8^{-\frac{1}{3}}$ (m) $\dfrac{1}{9}$

(f) 3^{-2} (n) 3^{-1}

4. Match the items:

(a) -2^{-1} (i) 3

(b) $3^{(-1)^{-1}}$ (j) 4

(c) $\dfrac{1^{-\frac{1}{2}}}{9}$ (k) $-\dfrac{1}{2}$

(d) $5^{(-1)^{-2}}$ (l) 5^3

(e) $\dfrac{1^{-2}}{2}$ (m) $\dfrac{1}{3}$

(f) $(5^{-3})^{-1}$ (n) 5

1. The matches are:

$$(a) \rightarrow (m)$$
$$(b) \rightarrow (n)$$
$$(c) \rightarrow (j)$$
$$(d) \rightarrow (i)$$
$$(e) \rightarrow (l)$$
$$(f) \rightarrow (k)$$

2. The matches are:

$$(a) \rightarrow (m)$$
$$(b) \rightarrow (l)$$
$$(c) \rightarrow (n)$$
$$(d) \rightarrow (j)$$
$$(e) \rightarrow (i)$$
$$(f) \rightarrow (k)$$

3. The matches are:

$$(a) \rightarrow (n)$$
$$(b) \rightarrow (k)$$
$$(c) \rightarrow (i)$$
$$(d) \rightarrow (j)$$
$$(e) \rightarrow (l)$$
$$(f) \rightarrow (m)$$

4. The matches are:

$$(a) \rightarrow (k)$$
$$(b) \rightarrow (m)$$
$$(c) \rightarrow (i)$$
$$(d) \rightarrow (n)$$
$$(e) \rightarrow (b)$$
$$(f) \rightarrow (l)$$

GEOMETRY QUIZ TWO

Pythagorean triples are integer numbers that satisfy the Pythagorean Theorem: $x^2 + y^2 = z^2$. If such a triple is found, then we can obtain another triple by multiplying each number by the same factor:

$$x^2 + y^2 = z^2$$

$$k^2 x^2 + k^2 y^2 = k^2 z^2$$

$$(kx)^2 + (ky)^2 = (kz)^2$$

The following triangles are right angle triangles. Fill in the missing values for the lengths of the sides. The figures are not to scale.

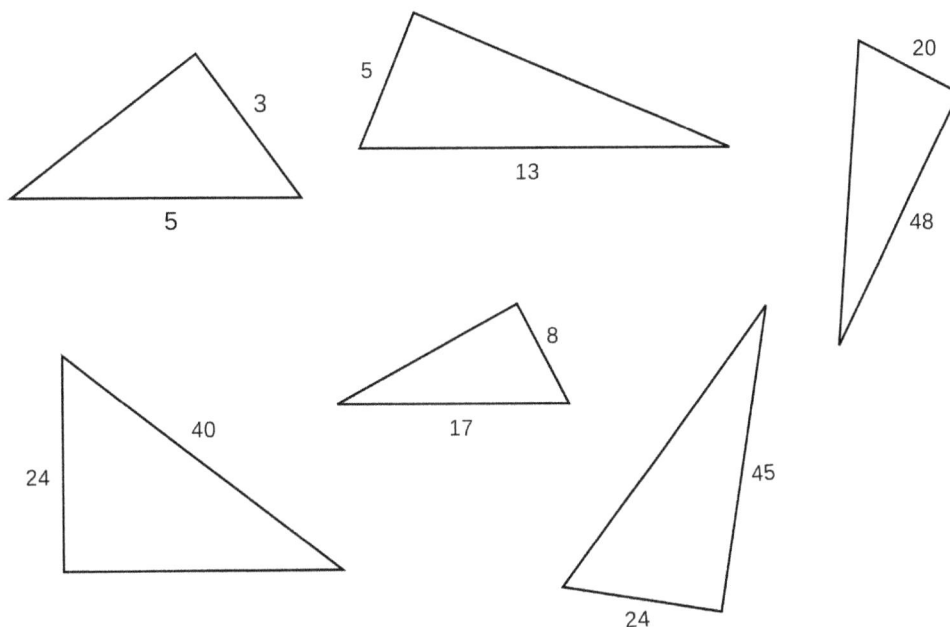

Solutions

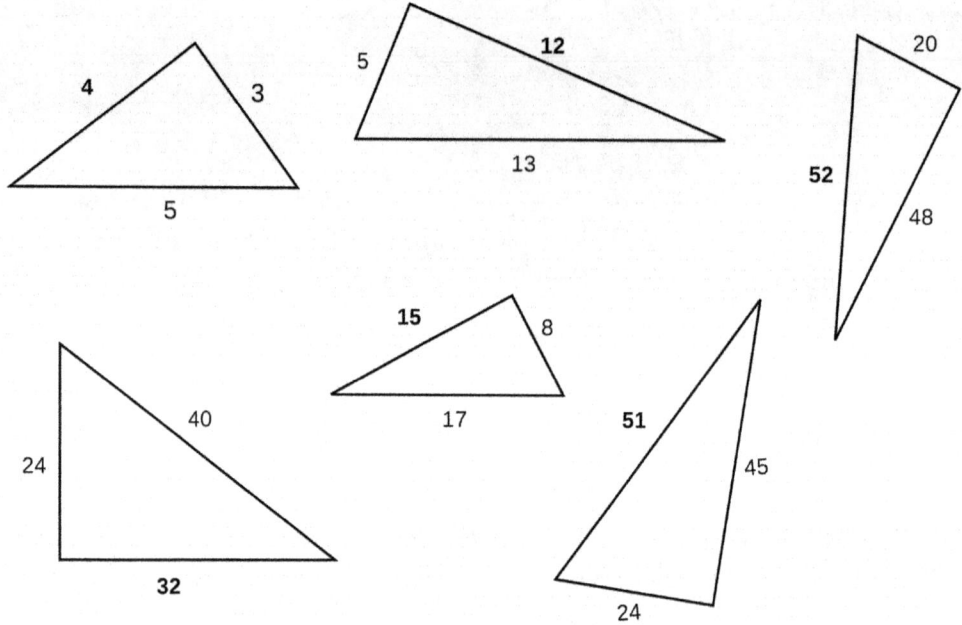

Memorize the Pythagorean triples that are bolded below! Their multiples should become easy to recognize in problems:

3:4:5, 6:8:10, 9:12:15, $\cdots$

5:12:13, 10:24:26, $\cdots$

8:15:17, 16:30:34, $\cdots$

7:24:25, 14:48:50, $\cdots$

Important note: Right triangles with integer side lengths **do not** also have integer degree angles! Special triangles $30° - 60° - 90°$ and $45° - 45° - 90°$ **do not** have integer length sides!

DATA ANALYSIS QUIZ TWO

1. We toss a coin and a die. What is the probability of getting a prime number on the die and a heads on the coin?

2. We toss three coins. What is the probability that two will fall heads and one will fall tail?

3. We have 6 cards, each with a different letter on it. How many different words can we make by using each card once?

4. We have a deck of cards. At least how many cards must we draw to be sure we hold two cards of the same suit?

5. People are voting for three candidates. If people are twice as likely to vote for A than for B, and four times as likely to vote for C than for A, what is the probability of voting for B?

6. Two pigs are eating two types of mushrooms: yellow and brown. Pig A eats yellow mushrooms with probability 0.6. Pig B eats brown mushrooms with probability 0.3. If they both eat the same number of mushrooms, what is the probability of a brown mushroom being eaten?

7. In how many ways can we color 5 circles of different sizes if we have 4 different colors to choose from?

8. Eight students play a game in two equal teams. In how many different ways can the teams be formed?

9. Only three athletes may be given the top three medals. If ties are allowed, how many possible ways of assigning medals are there?

10. If five horses enter a race, in how many ways can the first three horses be selected?

Answers to questions 1-10

1. Use the product rule:
$$\frac{1}{2} \cdot \frac{1}{2} = \frac{1}{4}$$

2. Use the sum and product rules:
$$3 \cdot \frac{1}{2} \cdot \frac{1}{2} \cdot \frac{1}{2} = \frac{3}{8}$$

3. $6! = 720$

4. 5 cards. Use the Pigeonhole Principle: *if we place n objects in $n - 1$ boxes, there is at least one box with two objects in it.* The suits are 4 boxes. If we draw five cards at random, there is at least one box with 2 cards in it.

5. Since $p(A) = 2p(B)$ and $p(C) = 4p(A)$, then $2p(B) + 8p(B) + p(B) = 11p(B) = 1$ and:
$$p(B) = \frac{1}{11}$$

6. Pig A eats brown mushrooms with probability $1 - 0.6 = 0.4$. If each pig eats M mushrooms, the probability of a brown mushroom being eaten is:
$$\frac{M \times 0.4 + M \times 0.3}{2 \times M} = \frac{0.7}{2} = 0.35 = 35\%$$

7. For each circle, there is a choice of 4 colors:
$$5^4 = 25 \times 25 = 625$$

8. Select 4 out of 8:
$$\binom{4}{8} = \frac{8!}{4! \cdot 4!} = \frac{5 \cdot 6 \cdot 7 \cdot 8}{1 \cdot 2 \cdot 3 \cdot 4} = 70$$

9. 1 way if all three get gold, 3 ways if one gets gold and two get silver, 3 ways if two get gold and one gets silver, 6 ways if gold, silver, and bronze are given out. There are 13 ways in total.

10. There are $_5C_3$ ways to select three horses. The three selected horses may arrive in $3!$ different possible orders. The total number of ways is:
$$\frac{5!}{3! \cdot 2!} \cdot 3! = 60$$

NUMBER SENSE QUIZ TWO

1. If we have to calculate $\dfrac{97}{101} \div \dfrac{11}{512}$ which of the following calculator entries are correct? (Check all that apply.)

(A) $97/101/512/11$

(B) $97 * 512/101/11$

(C) $97 * 512/(101 * 11)$

(D) $97/101 * (512/11)$

2. Calculate the percent change of:

(A) a 45% increase, followed by a 22% decrease

(B) a 20% increase, followed by a 20% increase, followed by a 20% increase

(C) a 20% increase, followed by a 10% decrease, followed by a 10% decrease

(D) a 4% decrease, followed by a 4% decrease, followed by a 50% increase

3. What fraction of a quantity is left if:

(A) We remove $\dfrac{3}{4}$, add $\dfrac{1}{4}$ to the remaining, and remove $\dfrac{1}{3}$ of the remaining.

(B) We add $\dfrac{1}{3}$, add $\dfrac{2}{3}$ to the remaining, and remove $\dfrac{3}{5}$ of the remaining.

(C) We remove $\dfrac{1}{5}$, remove $\dfrac{2}{3}$ from the remaining, and add $\dfrac{3}{4}$ to the remaining.

(D) We add $\dfrac{2}{3}$, add $\dfrac{1}{3}$ to the remaining, and remove $\dfrac{4}{7}$ of the remaining.

Answers to questions 1-3

1. Correct are (B), (C), and (D). Please actually input the numbers in the calculator to see the difference. The order of operations is left to right!

2. Assume a quantity of 100. The difference between the final quantity and 100 is the percent change:

(A) $100 * 1.45 * 0.78 = 113.1$ - a 13.1% increase

(B) $100 * 1.2 * 1.2 * 1.2 = 172.8$ - a 72.8% increase

(C) $100 * 1.2 * 0.9 * 0.9 = 97.2$ - a 2.8% decrease

(D) $100 * 0.96 * 0.96 * 1.5 = 138.24$ - a 38.24% increase

3. Assume a quantity of 1:

(A) $1 * \dfrac{1}{4} * \dfrac{5}{4} * \dfrac{2}{3} = \dfrac{10}{16 * 3} = \dfrac{5}{24}$

(B) $1 * \dfrac{4}{3} * \dfrac{5}{3} * \dfrac{2}{5} = \dfrac{40}{45} = \dfrac{8}{9}$

(C) $1 * \dfrac{4}{5} * \dfrac{1}{3} * \dfrac{7}{4} = \dfrac{28}{60} = \dfrac{7}{15}$

(D) $1 * \dfrac{5}{3} * \dfrac{4}{3} * \dfrac{3}{7} = \dfrac{20}{21}$

Diagnostic Three

This math section contains 18 questions to be solved in 25 minutes.

You are allowed to use a calculator. However, in order to minimize the time needed to complete the test, it is important to use the calculator as little as possible! Keep in mind:

- Entering operations in the calculator is more time consuming than performing the operations mentally.

- Data entry errors will be made in addition to other errors. Entering data is, in itself, a possible cause for error.

You can reduce the use of the calculator by memorizing well a short list of commonly used numbers: perfect squares from 1 to 20^2, powers of 2 from 2^0 to 2^{10}, some frequently used Pythagorean triples, etc. The list can be found in Appendix A, as well as on www.mathinee.com.

As you solve problems, build on the ability to recognize the numbers in this list. Every time you encounter them, tell yourself "hey, this is a power of 2", or "hey, this is a Pythagorean triple." After a short while, you will have memorized the list almost completely. But going over the list 'cold' a few times is also helpful, so go ahead and open Appendix A!

Before starting to solve, spend a few seconds to think about focusing, about making sure that you are ready to pay attention to every detail.

1. If $x = -1$ and $z = 2$ what is the value of the expression:

$$|x - z| + |z - x|$$

(A) 0 **(B)** 1 **(C)** 2
(D) 4 **(E)** 6

Ⓐ Ⓑ Ⓒ Ⓓ Ⓔ

2. Two parallel lines intersect a circle at 4 points. At least how many axes of symmetry has the figure thus formed?

(A) 0 **(B)** 1 **(C)** 2
(D) 4 **(E)** an infinity

Ⓐ Ⓑ Ⓒ Ⓓ Ⓔ

3. In the following data set, if the median value is 31 what is the mean?

$$\{13, 13, 28, 28, x, x, 58, 58\}$$

(A) 11.5 **(B)** 33.25 **(C)** 34.75

(D) 36 **(E)** 39

Ⓐ Ⓑ Ⓒ Ⓓ Ⓔ

4. The sum of five consecutive odd numbers is F. What is the difference between the largest and the smallest numbers in the sequence?

(A) 4

(B) 8

(C) $F - 4$

(D) $F - 10$

(E) cannot be determined

Ⓐ Ⓑ Ⓒ Ⓓ Ⓔ

5. If $\sqrt{4 - x} = 16$ then which of the following is true about $\sqrt{x}$:

(A) $\sqrt{x} = -252$
(B) $\sqrt{x} = 252^2$
(C) $\sqrt{x} = 12$
(D) $\sqrt{x} = -12$
(E) it is not real

Ⓐ Ⓑ Ⓒ Ⓓ Ⓔ

6. A survey of the number of show tickets purchased by families last year in the Sample County, has shown the following preferences:

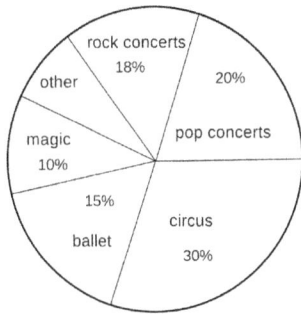

rock concerts 18%

other

20%

magic 10%

pop concerts

15%

ballet

circus 30%

If 49 families purchased tickets to shows that were not categorized, how many more families purchased tickets to rock concerts?

(A) 70

(B) 77

(C) 125

(D) 140

(E) 143

Ⓐ Ⓑ Ⓒ Ⓓ Ⓔ

7. What is the value of x that satisfies the equation:

$$\frac{3}{7} = 3 - \frac{1}{3} - \frac{1}{x}$$

(A) $-\dfrac{1}{3}$

(B) $\dfrac{15}{3}$

(C) $-\dfrac{15}{3}$

(D) $\dfrac{47}{21}$

(E) $\dfrac{21}{47}$

Ⓐ Ⓑ Ⓒ Ⓓ Ⓔ

8. In a mixture of gold and silver coins, there are 5 gold coins for every 7 silver coins. If 360 silver coins were removed, the ratio of gold coins to the total number of coins would become 3 : 4. How many coins are there in total?

(A) 30

(B) 90

(C) 180

(D) 810

(E) 1050

Ⓐ Ⓑ Ⓒ Ⓓ Ⓔ

9. What is the value of the expression:

$$\frac{1}{3} \cdot \frac{9}{4} \cdot \frac{8}{5} \cdot \frac{35}{3} \cdot \frac{1}{22}$$

(A) $\dfrac{7}{11}$

(B) 15.75

(C) $0.9\overline{54}$

(D) $\dfrac{7}{22}$

(E) 1

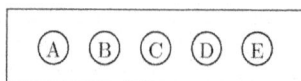

10. A point is selected at random in the circle with center O and radius 1. If the point ends up in the shaded area with probability $\dfrac{1}{\pi}$, what is the measure of the arc $\overarc{AB}$, in radians?

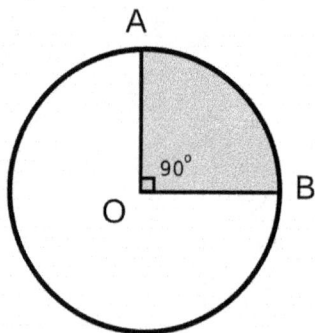

(A) 1

(B) 2

(C) $\dfrac{\pi}{3}$

(D) $\dfrac{\pi}{4}$

(E) cannot be determined

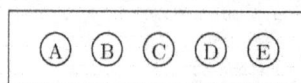

11. If $m^{\frac{1}{5}} = p^2$, then m is equivalent to:

(A) p^{10}

(B) p^{-3}

(C) $p^{\frac{2}{5}}$

(D) $p^{\frac{5}{2}}$

(E) $p^{-\frac{5}{2}}$

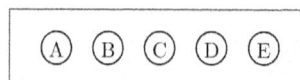

12. In the triangle ABC the point M divides the side AB in a ratio of $1 : 3$ and the point N divides the side BC in a ratio of $3 : 5$. What is the ratio between the shaded area and the area of the triangle ABC?

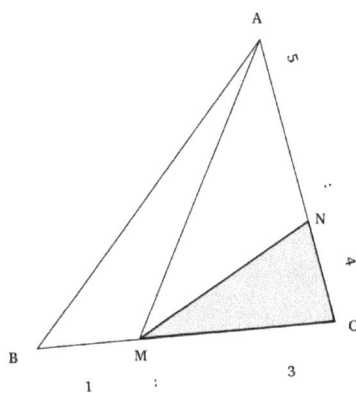

(A) $1 : 3$

(B) $1 : 4$

(C) $3 : 5$

(D) $3 : 4$

(E) $3 : 8$

Ⓐ Ⓑ Ⓒ Ⓓ Ⓔ

13. What is the sum of the digits of the number $10^{51} - 10$?

(A) 0

(B) 9

(C) 450

(D) 459

(E) $10^{51} - 1$

Ⓐ Ⓑ Ⓒ Ⓓ Ⓔ

14. An emergency room has seen 150 male and 140 female patients in one week. What is the probability that there was at least one day in which 20 or more female patients were seen?

(A) $\dfrac{2}{29}$

(B) $\dfrac{14}{15}$

(C) $\dfrac{4}{29}$

(D) $\dfrac{133}{29}$

(E) 1

Ⓐ Ⓑ Ⓒ Ⓓ Ⓔ

15. If the figure below represents the net of a box, which pair of points coincide?

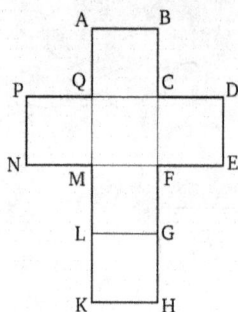

(A) (Q, C)

(B) (P, D)

(C) (B, G)

(D) (E, H)

(E) (N, L)

Ⓐ Ⓑ Ⓒ Ⓓ Ⓔ

16. A set of plastic chips consists of green and red chips only. If the probability of drawing a green chip blindfolded is $\dfrac{11}{13}$, what is the probability of drawing two chips with different colors from the whole set, without replacement?

(A) $\dfrac{121}{169}$

(B) $\dfrac{11}{78}$

(C) $\dfrac{22}{169}$

(D) $\dfrac{11}{39}$

(E) $\dfrac{11}{36}$

Ⓐ Ⓑ Ⓒ Ⓓ Ⓔ

14.1   Self-grade Diagnostic Three

Question	1	2	3	4	5	6	7	8
Level	1	1	1	2	2	2	2	3
Topic	A	G	S	N	A	S	A	
Answer	E	B	B	B	E	B	E	D
Correct								
Incorrect								
Skipped								

Question	9	10	11	12	13	14	15	16
Level	2	3	3	3	4	4	3	4
Topic	N	S	A	G	N	S	G	S
Answer	A	B	A	A	C	E	E	D
Correct								
Incorrect								
Skipped								

Are your incorrect answers:

mostly on questions in the first part of the test?	Yes	No
mostly on questions that have difficulty levels 3 and 4?	Yes	No
mostly from a specific topic (geometry, algebra, arithmetic)?	Yes	No
somewhat evenly spread throughout the test?	Yes	No
mostly due to minor errors?	Yes	No
mostly due to numeric errors?	Yes	No

A-algebra, N-number sense, G-geometry, S-statistics

14.2   Diagnostic Three Solutions

Question 1

The expression can be processed as follows:

$$\begin{aligned} |x-z|+|z-x| &= |-1-2|+|2-(-1)| \\ &= |-3|+|2+1| \\ &= 3+3 = 6 \end{aligned}$$

Note that: $2-(-1) = 2+1$. The absolute value leaves positive values unchanged but changes the sign of negative values.

Question 2

There is at least one axis of symmetry:

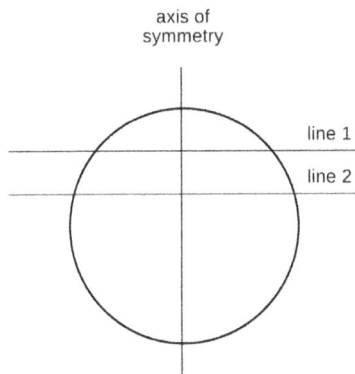

Question 3

There are 8 values in the set. Since 8 is even, the median value is the average of the two middle values:

$$\frac{28+x}{2} = 31$$

Solve for x to get $x = 34$. The data set is:

$$\{13, 13, 28, 28, 34, 34, 58, 58\}$$

and has the mean:

$$\frac{2 \times 13 + 2 \times 28 + 2 \times 34 + 2 \times 58}{8} = \frac{13 + 28 + 34 + 58}{4} = \frac{133}{4} = 33.25$$

Question 4

Two consecutive odd numbers differ by 2. If the smallest number is s, then the sequence is:

$$s, s+2, s+4, s+6, s+8$$

The difference between the largest and the smallest is 8.

Question 5

Since the quantity under the square root must be positive, $x \leq 4$. To obtain 16, x must be equal to -252:

$$\sqrt{4-x} = 16$$
$$4 - x = 256$$
$$x = -252$$

Since x is negative, there is no real value for $\sqrt{x}$.

Question 6

Calculate the percentage of tickets sold in the category "other":

$$100 - 18 - 20 - 30 - 15 - 10 = 7$$

There were $18 - 7 = 11\%$ more families who purchased "rock concert" tickets than tickets to "other".

If 49 families represent 7%, then 7 families represent 1%. Then, 11% represents 77 families.

Question 7

Solve for x:

$$\frac{3}{7} = \frac{9}{3} - \frac{1}{3} - \frac{1}{x}$$

$$= \frac{8}{3} - \frac{1}{x}$$

$$\frac{1}{x} = \frac{8}{3} - \frac{3}{7}$$

$$\frac{1}{x} = \frac{56}{21} - \frac{9}{21}$$

$$= \frac{56 - 9}{21} = \frac{47}{21}$$

$$x = \frac{21}{47}$$

Question 8

Denote the total number of coins by N. There are $\dfrac{7}{12}N$ silver coins. After removing 360 silver coins, the amount of silver coins becomes equal to $\dfrac{1}{4}(N - 360)$. We have the equation:

$$
\begin{aligned}
\frac{7}{12}N - 360 &= \frac{1}{4}(N - 360) \\[2mm]
\frac{7}{12}N - 360 &= \frac{1}{4}N - 90 \\[2mm]
\frac{7}{12}N - \frac{1}{4}N &= 360 - 90 \\[2mm]
N\left(\frac{7}{12} - \frac{1}{4}\right) &= 270 \\[2mm]
N\left(\frac{7 - 3}{12}\right) &= 270 \\[2mm]
N\frac{4}{12} &= 270 \\[2mm]
N &= \frac{270}{1} \times \frac{12}{4} = 270 \times 3 = 810
\end{aligned}
$$

Question 9

Simplify factors to get:

$$
\frac{1}{\not3} \cdot \frac{\not3 \cdot \not3}{\not4} \cdot \frac{\not4 \cdot \not2}{\not5} \cdot \frac{\not5 \cdot 7}{\not3} \cdot \frac{1}{\not2 \cdot 11} = \frac{7}{11}
$$

Question 10

The geometric probability is:

$$P = \frac{\text{favorable area}}{\text{total area}}$$

If the radius of the circle is R, the total area is πR^2.

The area of the shaded sector is:

$$A_{\text{sector}} = \frac{R^2 \times \theta}{2}$$

where θ is the measure, in radians, of the central angle that intercepts the arc $\overset{\frown}{AB}$. We can now solve for θ:

$$\frac{1}{\pi} = \frac{R^2 \times \theta}{2\pi R^2}$$

$$\frac{1}{\pi} = \frac{\theta}{2\pi}$$

$$1 = \frac{\theta}{2}$$

$$\theta = 2$$

Question 11

Raise both sides to the fifth power:

$$\left(m^{\frac{1}{5}}\right)^5 = \left(p^2\right)^5$$

and use the fact that, to calculate the power of a power, one has to multiply the exponents:

$$m^{\frac{1}{5} \cdot 5} = p^{2 \cdot 5}$$

$$m = p^{10}$$

Question 12

The altitudes from vertex A of the triangles ABC and AMC are equal in length. Therefore, the areas of the two triangles are in the same ratio as their bases:

$$\frac{A_{\triangle AMC}}{A_{\triangle ABC}} = \frac{3}{4}$$

Likewise, the areas of triangles AMC and MNC are in the same ratio as their bases AC and NC:

$$\frac{A_{\triangle MNC}}{A_{\triangle AMC}} = \frac{4}{9}$$

Therefore:

$$\frac{A_{\triangle AMC}}{A_{\triangle ABC}} \times \frac{A_{\triangle MNC}}{A_{\triangle AMC}} = \frac{3}{4} \times \frac{4}{9} = \frac{3}{9} = \frac{1}{3}$$

Question 13

The number $10^{51} - 10$ has 51 digits and consists of 50 digits of 9 and one digit of 0. The sum of its digits is $50 \times 9 = 450$.

Question 14

The words "at least" suggest that this is an application of the Pigeonhole Principle. Since 140 female patients were seen in a week, then there is at least one day in which 20 or more patients were seen. The probability is 1 since the event is certain.

Question 15

When folded into a cube, the sides MN and ML coincide. Their ends, points N and L become a single point.

Question 16

The probability of drawing a red chip blindfolded is:

$$1 - \frac{11}{13} = \frac{2}{13}$$

The probability of drawing two chips of different color by drawing without replacement is based on two possible events: *drawing a red chip followed by a green chip* OR *drawing a green chip followed by a red chip*. The combined probability is:

$$
\begin{aligned}
P(G, R) + P(R, G) &= \frac{11}{13} \cdot \frac{2}{12} + \frac{2}{13} \cdot \frac{11}{12} \\
&= \frac{11 \cdot \not{2}}{13 \cdot 6 \cdot \not{2}} \\
&= \frac{11}{13 \cdot 3} = \frac{11}{39}
\end{aligned}
$$

Algebra Quiz Three

1. For what real values x is it true that: $|x - 4| = x - 4$?

2. For what real values x is it true that: $|x - 4| = 4 - x$?

3. For what real values x is it true that: $x|x - 4| \leq 0$?

4. For what real values x is it true that: $x|x - 4| = 0$?

5. For how many real values x is it true that $x(x - 4) = -1$?

6. For what real values x is it true that $(x - 3)(5 - x) \geq 0$?

7. For how many real values x is it true that $x(5 - x) = -7$?

8. If $x < 0$ and $y > 0$, $x^3 \cdot y^3$ is
   always positive/always negative/sometimes positive.

9. If $x > 0$ and $y \leq 0$, then $x^2 \cdot y$ is
   always positive/negative/non-negative/non-positive.

10. If $x > 1$ and $y < 1$, then $\dfrac{y^2}{x}$ is
    always smaller/always greater/sometimes smaller than 1?

Answers to questions 1-10

1. Any $x \geq 4$.

2. Any $x \leq 4$.

3. $x \leq 0$ and $x = 4$.

4. $x = 0$ and $x = 4$.

5. It is a quadratic equation with positive discriminant. There are two real solutions.

6. $x \in [3, 5]$

7. It is a quadratic equation with negative discriminant ($\Delta = (-5)^2 - 4 \cdot (-7) = 25 + 28 = 53$) . There are no real solutions.

8. If $x < 0$, then $x^3 < 0$. Since $y^3 > 0$, the product $x^3 \cdot y^3$ is always negative.

9. $x^2 \cdot y$ is always non-positive. It can be zero if y is zero, and is otherwise negative.

10. $\dfrac{y^2}{x}$ is *sometimes* smaller than 1. If y is large and negative, the fraction can be larger than 1. Try, for example, the values $x = 2$ and $y = -5$. For other combinations of values, such as $x = 2$ and $y = 1/2$, the fraction is smaller than 1.

GEOMETRY QUIZ THREE

Special right triangles are triangles similar to triangles obtained by cutting a square in half and by cutting an equilateral triangle in half.

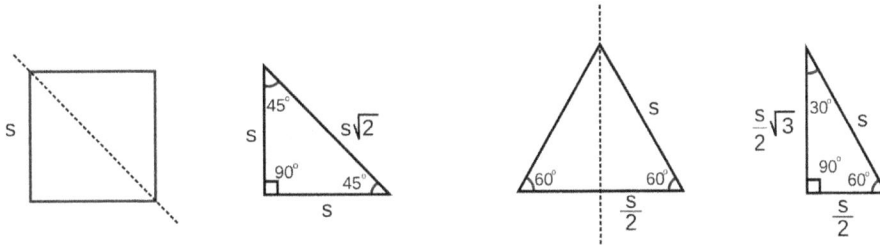

Write the missing side lengths and angle measures on the triangles in the figure (not to scale):

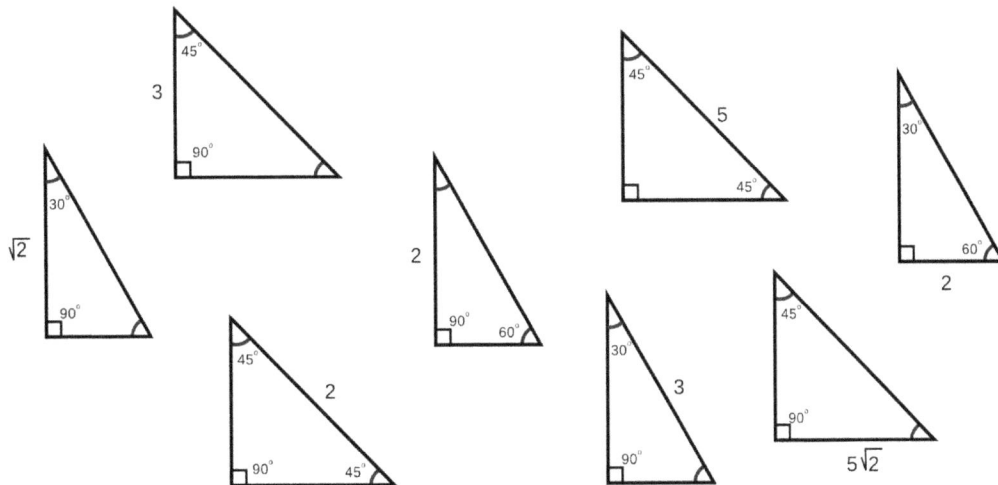

Solutions

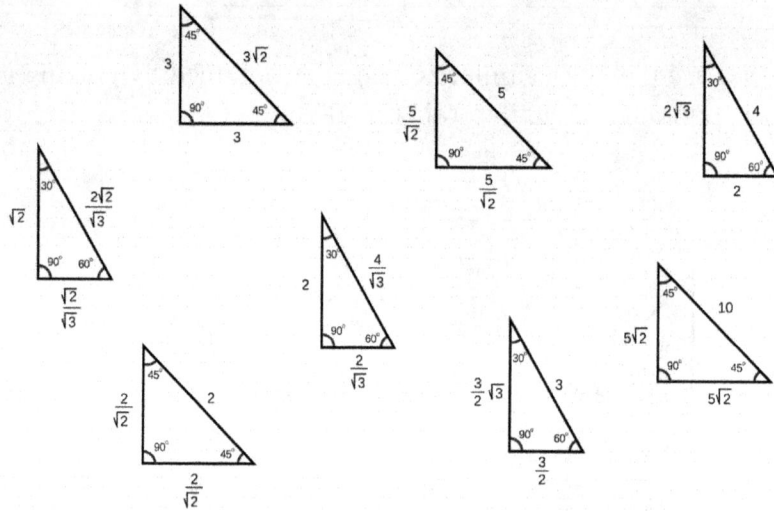

DATA ANALYSIS QUIZ THREE

1. How many ways can 5 identical crayons be placed in a row?

2. How many ways can 5 crayons of different colors be placed in a row?

3. How many ways can 5 crayons of 4 different colors be placed in a row?

4. How many ways can 5 crayons of 2 different colors be placed in a row?

5. Of 5 athletes who compete, 3 are going to obtain medals. If there are no ties possible, in how many ways can the medals be assigned?

6. If 5 people compete for two different jobs, in how many ways can the two openings be filled?

7. There are 3 choices for soup and 5 choices for the main course at Stella's Diner. Two fishermen stop by. In how many ways can they choose two completely different lunches, each consisting of soup and main course?

8. In how many ways can 5 bees visit two different kinds of flowers, if each bee visits a single flower?

9. If a person has the key to a locker room and does not know to which of the 26 rooms the key belongs, at most how many trials are needed until the match is found?

10. The probability of throwing two dice and obtain an even sum is larger/smaller/equal to the probability of obtaining an odd sum.

Answers to questions 1-10

1. 1 way.

2. $5! = 120$ ways.

3. There are 2 crayons of the same color:

$$\frac{5!}{2!} = 60$$

4. There are 2 crayons of a color and 3 crayons of the other color:

$$\frac{5!}{2! \cdot 3!} = 10$$

5. Choose 3 winners from the set of 5, then permute the three to obtain all possible prize assignments:

$$\frac{5!}{3! \cdot 2!} \cdot 3! = \frac{5!}{2!} = \frac{120}{2} = 60$$

6. Choose 2 of the 5 people, then multiply by 2 ways in which they can be assigned jobs.

$$\frac{5!}{3! \cdot 2!} \cdot 2 = 20$$

7. One fisherman has $3 \times 5 = 15$ choices, and the other has $2 \times 4 = 8$ choices that are different. The total number of combinations is $15 \times 8 = 120$.

8. In 6 possible ways:

Flower A	5 bees	4 bees	3 bees	2 bees	1 bee	0 bees
Flower B	0 bees	1 bees	2 bees	3 bees	4 bee	5 bees

9. At most 25 trials are needed. Note that the statement only requires to match the keys, not to also open the door.

10. The probability of throwing two dice and obtain an even sum is equal to the probability of obtaining an odd sum.

NUMBER SENSE QUIZ THREE

Statement	True	False	Sometimes true
Zero is neither a positive nor a negative integer.			
The product of an even number and an odd number is even.			
The result of raising a number to a negative power is negative.			
There is no maximum or minimum y coordinate of a point on a line.			
If the sum of the digits of a number is a multiple of 3, so is the number.			
2 is the only even prime number.			
Odd powers of negative numbers are positive.			
The area of a square in square inches is larger than the length of the side of the square, in inches.			
The reciprocal of a number is smaller than the number.			
Consecutive numbers are positive.			

Answers

Statement	True	False	Sometimes true
Zero is neither a positive nor a negative integer.	V		
The product of an even number and an odd number is even.	V		
The result of raising a number to a negative power is negative. Example: $(-3)^{-3} = -1/27$.			V
There is no maximum or minimum y coordinate of a point on a line.	V		
If the sum of the digits of a number is a multiple of 3, so is the number.	V		
2 is the only even prime number.	V		
Odd powers of negative numbers are positive.		V	
The area of a square in square inches is larger than the length of the side of the square, in inches.			V
The reciprocal of a number is smaller than the number.			V
Consecutive numbers are positive.			V

DIAGNOSTIC FOUR

This math section contains 18 questions to be solved in 25 minutes.

You are allowed to use a calculator. However, in order to minimize the time needed to complete the test, it is important to use the calculator as little as possible! Keep in mind:

- Entering operations in the calculator is more time consuming than performing the operations mentally.

- Data entry errors will be made in addition to other errors. Entering data is, in itself, a possible cause for error.

You can reduce the use of the calculator by memorizing well a short list of commonly used numbers: perfect squares from 1 to 20^2, powers of 2 from 2^0 to 2^{10}, some frequently used Pythagorean triples, etc. The list can be found in Appendix A, as well as on www.mathinee.com.

As you solve problems, build on the ability to recognize the numbers in this list. Every time you encounter them, tell yourself "hey, this is a power of 2", or "hey, this is a Pythagorean triple." After a short while, you will have memorized the list almost completely. But going over the list 'cold' a few times is also helpful, so go ahead and open Appendix A!

Before starting to solve, spend a few seconds to think about focusing, about making sure that you are ready to pay attention to every detail.

1. If $m = -5$ and $n = 3$, what is the value of the expression:

$$\frac{m \times (m + n)}{n \times 5}$$

(A) $\dfrac{50}{3}$ **(B)** 16.67 **(C)** 5

(D) $\dfrac{2}{3}$ **(E)** $\dfrac{-2}{3}$

Ⓐ Ⓑ Ⓒ Ⓓ Ⓔ

---

2. Find the measure of the angle x. (Figure not to scale.)

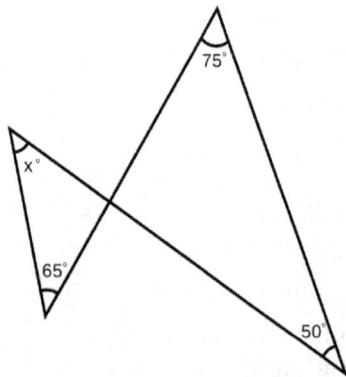

(A) 80° **(B)** 60° **(C)** 40°
(D) 50° **(E)** 85°

Ⓐ Ⓑ Ⓒ Ⓓ Ⓔ

3. In a set of blue and yellow cards, the probability of drawing a yellow card is exactly 0.88. The total number of cards cannot be:

(A) 20 **(B)** 25 **(C)** 50
(D) 75 **(E)** 100

Ⓐ Ⓑ Ⓒ Ⓓ Ⓔ

---

4. If 28% of the cars at a dealership are blue, and 79% of the cars are compact cars, what is the smallest possible percentage of blue compact cars?

(A) 0% **(B)** 7% **(C)** 28%
(D) 54% **(E)** 79%

Ⓐ Ⓑ Ⓒ Ⓓ Ⓔ

5. Which of the following could be the distance between the centers of two tangent circles with radii 8 and 3 units, respectively?

(A) 5 (B) 6 (C) 7
(D) 9 (E) 10

Ⓐ Ⓑ Ⓒ Ⓓ Ⓔ

6. 12^{50} is divisible by 2^k. What is the largest integer k with this property?

(A) 2 (B) 4 (C) 50
(D) 100 (E) 200

Ⓐ Ⓑ Ⓒ Ⓓ Ⓔ

7. The interior angles of a regular pentagon (polygon with 5 sides) are equal to $108°$. A figure is constructed by connecting the midpoints of each pair of neighboring sides. What is the measure of the interior angles of this figure?

(A) $54°$ (B) $100°$ (C) $108°$
(D) $120°$ (E) $216°$

Ⓐ Ⓑ Ⓒ Ⓓ Ⓔ

8. If the probability of drawing a red chip out of a bag with red and yellow chips is $\frac{3}{5}$, which of the following is the ratio of the number of red chips to the number of yellow chips?

(A) $3:2$ (B) $2:3$ (C) $5:3$
(D) $3:8$ (E) $3:5$

Ⓐ Ⓑ Ⓒ Ⓓ Ⓔ

9. The results of a customer survey have been summarized in a bar chart. What fraction of the responses had 4 or 5-star ratings?

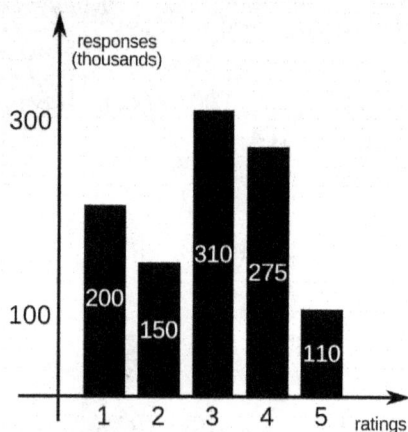

(A) 0.36

(B) $\dfrac{7}{19}$

(C) $\dfrac{1}{3}$

(D) $\dfrac{2}{5}$

(E) $\dfrac{7}{12}$

Ⓐ Ⓑ Ⓒ Ⓓ Ⓔ

10. A rectangle with side lengths 6π and 8π is inscribed in a circle (all the vertices of the rectangle are on the circle.) What is the area of the circle?

(A) 25π

(B) $25\pi^2$

(C) $25\pi^3$

(D) $100\pi^2$

(E) $100\pi^3$

Ⓐ Ⓑ Ⓒ Ⓓ Ⓔ

11. Which of the following numbers has the smallest reciprocal?

(A) -1

(B) $\dfrac{1}{3}$

(C) $\dfrac{1}{\pi}$

(D) π

(E) 3

Ⓐ Ⓑ Ⓒ Ⓓ Ⓔ

12. 20 cows yield 100 gallons of milk per day. 60 calves drink 100 gallons of milk per day. How many days will it take 50 cows to produce the milk needed by 75 calves over 10 days?

(A) 2

(B) 4

(C) 5

(D) 6

(E) 10

Ⓐ Ⓑ Ⓒ Ⓓ Ⓔ

13. If $(t-2)^2 = t$ and $t \neq 1$, what is the value of $(t-2)(t+2)$?

(A) 0

(B) 1

(C) 6

(D) 12

(E) 16

Ⓐ Ⓑ Ⓒ Ⓓ Ⓔ

14. The table summarizes the outcomes of spinning a spinner. Which of the spinners A through E is the most likely to have been used?

Clover	120
Daisy	123
Thyme	244
Mint	479

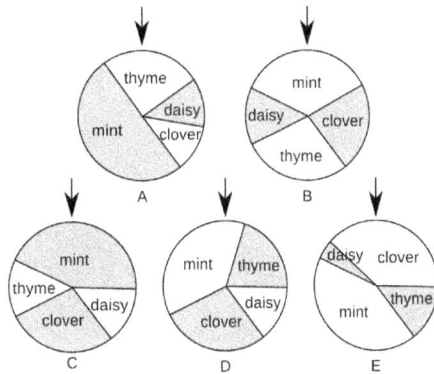

Ⓐ Ⓑ Ⓒ Ⓓ Ⓔ

15. Two parallel lines intercept two concurrent lines as in the figure.

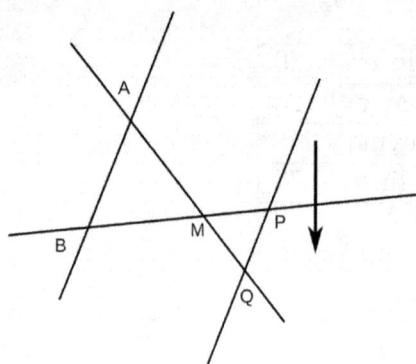

By moving the line that passes through P and Q downward, the ratio of the areas of triangles ABM and MPQ:

(A) increases

(B) decreases

(C) remains the same

(D) increases, then decreases

(E) decreases, then increases

Ⓐ Ⓑ Ⓒ Ⓓ Ⓔ

---

16. What is the smallest product one can get by multiplying two numbers from the set:

$$\left\{-\frac{1}{10}, 0, 1, \frac{2}{3}, 0.75\right\}$$

(A) $-\dfrac{1}{10}$

(B) $-\dfrac{2}{30}$

(C) $-\dfrac{9}{120}$

(D) -0.075

(E) 0

Ⓐ Ⓑ Ⓒ Ⓓ Ⓔ

---

17. If $(a+b)^3 < a^3 + b^3$ then how many of the following statements may be true?

i. $a < 0, b > 0$

ii. $a < 0, b < 0$

iii. $a > 0, b > 0$

iv. $a \leq 0, b > 0$

v. $a < b < 0$

(A) 0

(B) 1

(C) 2

(D) 3

(E) 4

Ⓐ Ⓑ Ⓒ Ⓓ Ⓔ

18. What should the value of k be if we want the solution of the following system to be a line:

$$13x + 5y = 23$$
$$kx + 35y = 161$$

(A) 13
(B) 18
(C) 26
(D) 91
(E) 93

Ⓐ Ⓑ Ⓒ Ⓓ Ⓔ

19. A truck 8 ft wide is stopped in the middle of a street that is 36 ft wide. What is the distance between the curb and the side of the truck?

(A) 10
(B) 12
(C) 14
(D) 15
(E) 16

Ⓐ Ⓑ Ⓒ Ⓓ Ⓔ

20. In a group of people, each person has 6 friends. How many people are there in the group?

(A) 6
(B) 7
(C) 12
(D) 36
(E) 64

Ⓐ Ⓑ Ⓒ Ⓓ Ⓔ

19.1   Self-grade Diagnostic Four

Question	1	2	3	4	5	6	7	8	9	10
Level	1	1	1	1	2	2	2	2	2	2
Topic	N	G	S	N	G	N	G	S	S	G
Answer	D	B	A	B	A	D	C	A	B	C
Correct										
Incorrect										
Skipped										

Question	11	12	13	14	15	16	17	18	19	20
Level	2	3	3	2	4	3	3	4	3	3
Topic	N	N	A	S	G	N	A	A	N	N
Answer	A	C	D	A	B	A	E	D	C	B
Correct										
Incorrect										
Skipped										

Are your incorrect answers:

mostly on questions in the first part of the test?	Yes	No
mostly on questions that have difficulty levels 3 and 4?	Yes	No
mostly from a specific topic (geometry, algebra, arithmetic)?	Yes	No
somewhat evenly spread throughout the test?	Yes	No
mostly due to minor errors?	Yes	No
mostly due to numeric errors?	Yes	No

A-algebra, N-number sense, G-geometry, S-statistics

19.2 Diagnostic Four Solutions

Question 1 Substitute. Apply the order of operations and the rule of signs:

$$\frac{-5 \cdot (-5 + 3)}{3 \times 5} = \frac{-5 \times -2}{3 \times 5} = \frac{2}{3}$$

The incorrect answer choices represent various frequent calculator errors.

Question 2

Use the facts:

- The interior angles of a triangle add up to $180°$.

- Vertical angles have the same measure.

Simply put, since the unmarked angles are vertical and have the same measure, the other two pairs of angles must have the same sum:

$$65 + x = 75 + 50$$

and $x = 60°$.

Question 3

Since the number of cards must be an integer, there can be 25 cards or any multiple thereof:

$$\frac{88}{100} = \frac{22}{25} = \frac{44}{50} = \frac{66}{75}$$

20 is not a multiple of 25 and, therefore, cannot be the total number of cards.

Question 4

Since $28 + 79 = 107$, there are at least 7% blue compact cars.

Question 5

The distance between the centers of two tangent circles can only be:

- the sum of the radii - if the circles are tangent exterior;
- the positive difference of the radii - if the circles are tangent interior.

For the circles specified, the distance can only be 5 or 11 units. Only 5 is an answer choice. The circles must be tangent interior.

Question 6

Since:

$$12^{50} = (2^2 \cdot 3)^{50} = 2^{100} \cdot 3^{50}$$

The largest k is 100.

Question 7

The figure formed is also a regular pentagon. As such, its interior angles have a measure of 108°.

Question 8

The probability is:

$$P(\text{red}) = \frac{\text{Number of red chips}}{\text{Total number of chips}} = \frac{3}{5}$$

Because:

$$\frac{3}{5} = \frac{3x}{5x} = \frac{3x}{3x + 2x}$$

it follows that, if the number of red chips is of the form $3x$, the number of yellow chips must be of the form $2x$. The ratio of red to yellow is: $3 : 2$.

Question 9

Since both the numerator and the denominator of the fraction are thousands, we can just use the number of thousands. The number of responses with 5-star ratings is 110 and the number of responses with 4-star ratings is 275 - in total $110 + 275 = 385$ responses. The total number of responses is:

$$200 + 150 + 310 + 275 + 110 = 1045$$

The fraction is:

$$\frac{385}{1045} = \frac{5 \cdot 7 \cdot 11}{5 \cdot 11 \cdot 19} = \frac{7}{19}$$

Question 10

The diameter of the circle forms a right angle triangle with the sides of the rectangle. The sides are in a Pythagorean ratio of $3 : 4 : 5$:

$$6\pi, \ 8\pi, 10\pi$$

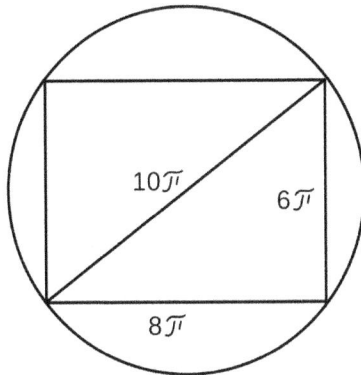

Therefore, the radius of the circle is 5π and its area is:

$$A_{\text{circle}} = \pi R^2 = \pi (5\pi)^2 = 25\pi^3$$

Question 11

The reciprocal of a number has the same sign as the number. Since negative numbers are smaller than positive numbers, the smallest reciprocal is $-1 = \dfrac{1}{-1}$.

Question 12

Start by how much milk is needed.

- 6 calves drink 10 gallons a day
- 3 calves drink 5 gallons a day
- 75 calves drink 125 gallons a day
- 75 calves drink 1250 gallons in 10 days

Now see how many cows you need:

- 1 cow yields 5 gallons a day
- 50 cows yield 250 gallons a day
- 1250 gallons are produced in 5 days

Question 13

The equation is quadratic and must have 0, 1, or 2 real roots. Notice that $t = 4$ is a root other than $t = 1$. Then $(t - 2)(t + 2) = t^2 - 4 = 16 - 4 = 12$. If you do not notice, solve the quadratic equation:

$$t^2 - 5t + 4 = 0$$

Question 14

In the table of outcomes, the frequency of the thyme is almost double that of the daisy. Also, the frequency of mint is almost double that of thyme. The only spinner that matches is (A).

Question 15

The line has positive slope. As it moves downwards, the points P and Q move towards the right. The area of triangle MPQ increases. The ratio of the areas of ABM and MPQ in this order, decreases (since numerator stays unchanged while the denominator increases.)

Question 16

The smallest product must be negative. The smallest negative product is the product with the largest absolute value: -0.1.

Question 17

The inequality yields:

$$
\begin{aligned}
(a+b)^3 \;=\; a^3 + 3a^2b + 3ab^2 + b^3 &< a^2 + b^3 \\
3a^2b + 3ab^2 &< 0 \\
2ab(a+b) &< 0 \\
ab(a+b) &< 0
\end{aligned}
$$

The product is negative if:

(α) one of the three factors is negative and two are positive

(β) all three factors are negative

    i. If a and b have different signs, it is possible for the sum $a+b$ to be positive, in which case condition (α) is fulfilled.

    ii. If a and b are both negative, their sum is negative, and condition (β) is fulfilled.

    iii. If a and b are both positive, all three factors are positive and none of the conditions is fulfilled.

    iv. This is similar to **i.**.

    v. This is similar to **ii.**.

Therefore, 4 of the statements may be true, depending on the actual values of a and b.

Question 18

If $k = 91$, the two equations are identical:

$$13 \times 7x + 5 \times 7y = 23 \times 7$$

They both represent the same line.

Question 19

The width not occupied by the truck is $36 - 8 = 28$ ft. Divide this by 2 to find the distance between the side of the truck and the curb: $28 \div 2 = 14$ ft.

Question 20

If each person has 6 friends, there must be 7 persons in total.

Algebra Quiz Four

1. If $a > 0$, $b > 0$, and $c < 0$, what can be the sign of each expression below:

Expression	+	−	Either
abc			
$ab(c - 2)$			
$(a - 2)bc$			
$a^5 b^3 c^7$			
$(a - 3)^2 b(c + 4)^4$			

2. If $a > 1$, and $b < 1$, how does each expression compare to 1?

Expression	> 1	< 1	Either
ab			
$a + b$			
$a - b$			
$b + \dfrac{1}{a}$			
$b \times \dfrac{1}{a}$			

Answers to questions 1-2

1. Use appropriate numeric values to experiment.

Expression	> 0	< 0	Either
abc		V	
$ab(c-2)$		V	
$(a-2)bc$			V
$a^5b^3c^7$		V	
$(a-3)^2b(c+4)^4$	V		

2. Use appropriate numeric values to experiment.

Expression	> 1	< 1	Either
ab			V
$a+b$	V		
$a-b$			V
$b + \dfrac{1}{a}$			V
$b \times \dfrac{1}{a}$		V	

GEOMETRY QUIZ FOUR

Which of the following sets of lengths can be sides of a non-degenerate triangle? (Hint: apply the triangle inequalities. All three must be valid.)

Note: A degenerate triangle is a triangle that has a vertex placed on the side opposite to it and has, in fact, become a segment. Identify degenerate triangles among the ones in the table. Degenerate triangles are also called *triangles of area zero*.

Set of lengths	Yes	No	Degenerate
1, 2, 3			
11, 12, 13			
4, 6, 10			
15, 20, 11			
15, 15, 30			
1, 100, 103			
42, 53, 10			
43, 53, 11			
3a, 4a, 7a			

Answers to questions

Set of lengths	Yes	No	Degenerate
1, 2, 3		V	Yes
11, 12, 13	V		
4, 6, 10		V	Yes
15, 20, 11	V		
15, 15, 30		V	Yes
1, 100, 103		V	
42, 53, 10		V	
43, 53, 11	V		
3a, 4a, 7a		V	Yes

DATA ANALYSIS QUIZ FOUR

Use the data set $\{3, 5, 10, 14\}$ to practice averages. Problems 1-5 refer to this data set.

1. What value should we add to the set to increase its average by 1?

2. What value should we add to the set to decrease its average by 1?

3. What value should we add to the set to increase its average by 10%?

4. What value should we add to the set to decrease its average by 10%?

5. By how much percent does the average increase if we add the value 16 to the set?

6. Each of the values in a data set with 100 points increases by 4. By how much does the average increase?

7. Each of the values in a data set with 100 points increases 4 times. By how much does the average increase?

8. A data set of 8 values has an average of 92. Two of the values have been lost and the set looks like: $\{x, y, 90, 94, 94, 100, 104, 104\}$. What is the average of x and y?

9. The average of x, y, and z is 50 and the average of a and b is 98. What is the average of x, y, a, b, and z?

10. What is the average of the first five positive perfect squares?

Answers to questions 1-10

1. If the average is 9 and there are 5 data points, the sum of all the values is $9 \times 5 = 45$. Since $45 - 32 = 13$, we must add the value 13 to the set.

2. If the average is 7 and there are 5 data points, the sum of all the values is $7 \times 5 = 35$. Since $35 - 32 = 3$, we must add the value 3 to the set.

3. If the average is 8.8 and there are 5 data points, the sum of all the values is $8.8 \times 5 = 44$. The value $44 - 32 = 12$ must be added to the set.

4. If the average is 7.2 and there are 5 data points, the sum of all the values is $7.2 \times 5 = 36$. The value $36 - 32 = 4$ must be added to the set.

5. The sum of the values becomes $32 + 16 = 48$. There are 5 data points and the average is $48 \div 5 = 9.6$. The average has increased by $1.6 \div 8 = 20\%$.

6. Denote the average with A. The sum of the values is $100A$. If the sum of the values increases by 400, then the new average becomes:
$$A_{\text{new}} = \frac{100A + 400}{100} = A + 4$$

7. Denote the average with A. The sum of the values is $100A$. If each value quadruples, then she sum quadruples and becomes $400A$, then the new average becomes:
$$A_{\text{new}} = \frac{400A}{100} = 4 \times A$$

8. The sum of the known values is 586 and the total sum of the values in the data set is $92 \times 8 = 736$. The difference is the sum $x + y = 150$. The average of x and y is 75.

9. $x + y + z = 150$ and $a + b = 196$. $x + y + z + a + b = 346$. The average is $346 \div 5 = 69.2$.

10. $(1 + 4 + 9 + 16 + 25) \div 5 = 11$

NUMBER SENSE QUIZ FOUR

1. Find the values of x and y in each identity. Check with the answer sheet on the verso and give yourself one point for each correct answer.

Exponential identities	x	y	Your score
$5.6 \times 10^{-5} = 560 \times 10^x = 0.056 \times 10^y$			
$110 \times 10^{10} = 0.11 \times 10^x = 11000 \times 10^y$			
$15.06 \times 10^{-8} = 1.506 \times 10^x = 1506 \times 10^y$			
$200 \times 10^9 = 2 \times 10^x = 0.0002 \times 10^y$			
$2013 \times 10^6 = 2.013 \times 10^x = 201.3 \times 10^y$			
$10101 \times 10^{-4} = 1.0101 \times 10^x = 10101000 \times 10^y$			
$0.005 \times 10^{-9} = 5 \times 10^x = 5000 \times 10^y$			
$4.051 \times 10^4 = 4051 \times 10^x = 0.0004051 \times 10^y$			
$10,000 \times 10^{-11} = 10^x = 10^{-y}$			
$304 \times 10^{-7} = 3.04 \times 10^{-x} = 3040 \times 10^{-y}$			
$78 \times 10^8 = 7.8 \times 10^{-x} = 78,000 \times 10^{-y}$			

Answers to question 1

Exponential identities	x	y	Your score
$5.6 \times 10^{-5} = 560 \times 10^x = 0.056 \times 10^y$	-7	-3	
$110 \times 10^{10} = 0.11 \times 10^x = 11000 \times 10^y$	13	8	
$15.06 \times 10^{-8} = 1.506 \times 10^x = 1506 \times 10^y$	-7	-10	
$200 \times 10^9 = 2 \times 10^x = 0.0002 \times 10^y$	11	15	
$2013 \times 10^6 = 2.013 \times 10^x = 201.3 \times 10^y$	9	7	
$10101 \times 10^{-4} = 1.0101 \times 10^x = 10101000 \times 10^y$	0	-7	
$0.005 \times 10^{-9} = 5 \times 10^x = 5000 \times 10^y$	-12	-15	
$4.051 \times 10^4 = 4051 \times 10^x = 0.0004051 \times 10^y$	1	8	
$10,000 \times 10^{-11} = 10^x = 10^{-y}$	-8	8	
$304 \times 10^{-7} = 3.04 \times 10^{-x} = 3040 \times 10^{-y}$	5	8	
$78 \times 10^8 = 7.8 \times 10^{-x} = 78,000 \times 10^{-y}$	-9	-5	

DIAGNOSTIC FIVE

This math section contains 18 questions to be solved in 25 minutes.

You are allowed to use a calculator.

The first 8 questions are multiple choice, the remaining 10 questions are student response questions. For the student response questions, the answer consists of 4 characters chosen from the symbols: . /, and any of the ten digits. The answer may not start with a zero. Answers that have both a fractional and a decimal representation may be represented either way. For example, the representations:

$$1/4$$

and

$$.25$$

are equivalent. Note that 0.25 is not a possible entry.

Before starting to solve, spend a few seconds to think about focusing, about making sure that you are ready to pay attention to every detail.

1. If the operation $\oslash$ is defined as:

$$a \oslash b = \frac{a+1}{b+1}$$

what is the value of $1 \oslash 2 \oslash 1$?

(A) 2

(B) $\frac{4}{5}$

(C) $\frac{5}{3}$

(D) $\frac{5}{6}$

(E) $\frac{5}{4}$

Ⓐ Ⓑ Ⓒ Ⓓ Ⓔ

2. In a warehouse there are 760 items, of which half are barcoded and half are not. More newly purchased items which do not have a barcode are added until the barcoded items form 20% of the items that are not barcoded. How many items are there now in total?

(A) 152 (B) 608 (C) 1520
(D) 1900 (E) 2280

Ⓐ Ⓑ Ⓒ Ⓓ Ⓔ

3. A $5 \times 8 \times 3$ box is partially filled with liquid. If it is $\frac{2}{5}$ full when the longest side is up, to what fraction of its volume is it filled when the shortest side is up?

(A) $\frac{3}{5}$

(B) $\frac{8}{5}$

(C) $\frac{2}{15}$

(D) $\frac{8}{15}$

(E) $\frac{2}{5}$

Ⓐ Ⓑ Ⓒ Ⓓ Ⓔ

4. A package travels by air for 800 miles in 2 hours and by truck for 100 miles in 1.5 hours. Waiting times, while in transit, add up to 5.5 hours. What is the average transit speed of the package from origin to destination, in miles per hour?

(A) 100 (B) 164 (C) 360
(D) 450 (E) 900

Ⓐ Ⓑ Ⓒ Ⓓ Ⓔ

5. If the coordinates of point A are $(-5, k)$ and the coordinates of point B are $(5, -10)$, what value must k have for the two points to be symmetric with respect to the y-axis:

(A) -5

(B) -10

(C) 0

(D) 5

(E) 10

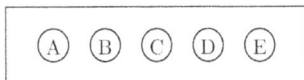

6. A pound of tangerines costs $4p$ dollars. Two pounds of apples cost $4a$ dollars. What is the price per pound, in dollars, of a mixture of equal parts of tangerines and apples?

(A) $2p + a$

(B) $p + a$

(C) $a + 2p$

(D) $8p + 4a$

(E) $4p + 2a$

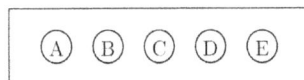

7. If today is Thursday, what day of the week will it be 125 days from now?

(A) Thursday

(B) Friday

(C) Saturday

(D) Tuesday

(E) Wednesday

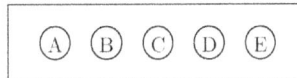

8. Seven eighths of the volume of a square pyramid with height 1 unit are filled with liquid. If the pyramid is placed to rest with the square base on a flat table, what height does the liquid reach to?

(A) 0.125

(B) 0.2

(C) 0.5

(D) 0.75

(E) 0.8

9. What is the value x that makes the mean and the median of the following data set equal?

$$\{2,\ 11,\ x,\ 20,\ 24,\ 35\}$$

10. A bag contains 40 cards numbered from 2 to 41. If we remove a card at random, what is the probability that the number on it is a multiple of 5 or of 4?

11. $S = \{-1, 0, 3\}$ and $T = \{-3, 0, 1\}$. If s is in S and t is in T, how many different values are there for the product $s \cdot t$?

12.   In a solution of 36% alcohol in water, what fraction of the water is the alcohol?

13. If $2a + 3b$ is 130% of $5b$, what is the ratio $a : b$?

14. If the side of each square in the cartesian grid below has length 1 unit, how many of the distances between distinct pairs of points from the set $\{A, B, C, D, E\}$ are rational numbers?

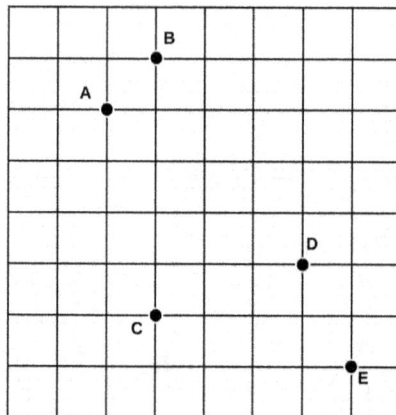

15. In the isosceles triangle OAB with $OA = AB$, M is the midpoint of OB. Extend AM to point C so that the area of $\triangle OMC$ twice as large as the area of $\triangle OAB$. The coordinates of point A are $(d, -7)$. What is the y-coordinate of point C?

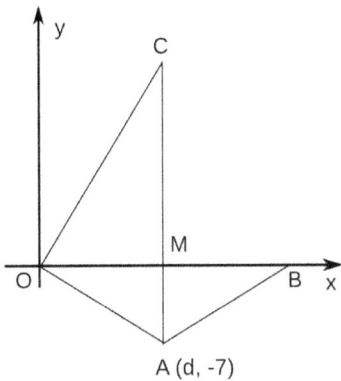

16. If the average A of 5 scores decreases by 10% when a new score of 188 is added to the set of scores, what is the value of A?

17. In the sequence of integer numbers:

$$a, \ 2a, \ 2a - 3, \ 4a - 6, \ 4a - 9, \ \cdots$$

each even term is the double of the preceding term and each odd term is lower than the preceding term by 3. What is the value of a for which the sequence is periodic?

18. A line with negative slope forms a triangle with the origin of the system of coordinates, like in the figure. The coordinates of point P are $(11, 0)$. By how many units must the line be translated in the positive x-direction for the area of the triangle to become 9 times larger?

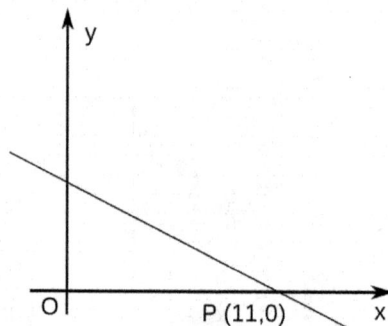

P (11,0)

24.1   Self-grade Diagnostic Five

Question	1	2	3	4	5	6	7	8
Level	1	1	1	1	2	2	2	3
Topic	A	N	G	N	G	A	N	G
Answer	D	E	E	A	B	A	D	C
Correct								
Incorrect								
Skipped								

Question	9	10	11	12	13	14	15	16	17	18
Level	2	2	2	3	3	3	4	4	4	4
Topic	S	S	N	N	N	A	G	S	A	G
Answer	16	2/5	4	9/16	7 : 4	3	28	470	3	22
Correct										
Incorrect										
Skipped										

Are your incorrect answers:

mostly on questions in the first part of the test?	Yes	No
mostly on questions that have difficulty levels 3 and 4?	Yes	No
mostly from a specific topic (geometry, algebra, arithmetic)?	Yes	No
somewhat evenly spread throughout the test?	Yes	No
mostly due to minor errors?	Yes	No
mostly due to numeric errors?	Yes	No

A-algebra, N-number sense, G-geometry, S-statistics

24.2 Diagnostic Five Solutions

Question 1 Process the expression from left to right:

$$1 \oslash 2 = \frac{1+1}{2+1} = \frac{2}{3}$$

$$\frac{2}{3} \oslash 1 = \frac{\frac{2}{3}+1}{1+1} = \frac{\frac{5}{3}}{2} = \frac{5}{6}$$

Question 2

380 items are barcoded and 380 are not barcoded. After adding x more items that are not barcoded, the number of items increases to $380 + x$. The barcoded items represent 20% of this number:

$$\begin{aligned}
380 &= \frac{20}{100}(380 + x) \\
380 &= \frac{1}{5}(380 + x) \\
5 \cdot 380 &= 380 + x \\
x &= 5 \cdot 380 - 380 \\
x &= 4 \cdot 380 = 1520
\end{aligned}$$

Therefore, the total number of items is $1520 + 760 = 2280$.

Question 3

The same fraction of the total volume will be occupied.

Question 4

The average speed is equal to the total distance divided by the total time:

$$S = \frac{800 + 100}{2 + 1.5 + 5.5} = \frac{900}{9} = 100 \text{ mph}$$

Question 5

To be symmetric with respect to the y-axis the points must be on a line perperdicular to the axis, at the same perpendicular distance from the axis and on opposite sides of it. The points $(-5, -10)$ and $(5, -10)$ are symmetric with respect to the y-axis.

Question 6

Two lbs of tangerines cost $2 \times 4p = 8p$ dollars. Two lbs of apples and two lbs of tangerines cost $8p + 4a$. Divide this by 4 lbs to obtain the price per pound: $2p + a$.

Question 7

Divide 125 by 7 and find the remainder.

$$125 = 7 \times 17 + 6$$

There are 17 complete weeks and 6 extra days. Count the extra days: Thursday, Friday, Saturday, Sunday, Monday, Tuesday.

Question 8

The volume of the pyramid is:

$$V_{\text{pyr}} = \frac{A_{\text{base}} \cdot H}{3}$$

The empty part of the pyramid is a pyramid similar to the larger one. The volume of the empty part of the pyramid is:

$$V_{\text{empty}} = \frac{1}{8} \times V_{\text{pyr}} = f^3 \times V_{\text{pyr}}$$

where f is a scaling factor. From here:

$$f^3 = \frac{1}{8}$$

$$f = \frac{1}{2}$$

From similarity, it follows that the height of the empty part is equal to one half of the height of the pyramid. The height reached by the water level is 0.5 units.

Question 9

Set up the equation:

$$\frac{2 + 11 + x + 20 + 24 + 35}{6} = x + 202$$

and solve for x:

$$\frac{92 + x}{6} = \frac{20 + x}{2}$$

$$92 + x = 60 + 3x$$

$$32 = 2x$$

$$16 = 6$$

Question 10

There are 8 multiples of 5 from 2 to 41. There are 10 multiples of 4 from 2 to 41. However, two of the multiples are the same in both sets: 20 and 40. There number of favorable outcomes is: $8 + 10 - 2 = 16$. The probability is:

$$\frac{16}{40} = \frac{2}{5}$$

Question 11

The products are:

$$-1 \cdot (-3) = 3$$

$$-1 \cdot 1 = -1$$

$$0 \cdot \text{any} = 0$$

$$3 \cdot (-3) = -9$$

$$3 \cdot 1 = 3$$

The different values are: $\{-9, -1, 0, 3\}$. In total, there are 4 different values.

Question 12

Denote the total volume of the mixture by M and the required fraction by f:

$$\frac{36}{100}M = f \cdot \frac{64}{100}M$$
$$36 = f \cdot 64$$
$$f = \frac{36}{64} = \frac{9}{16}$$

Question 13

Set up an equation for $2a + 3b$:

$$2a + 3b = \frac{130}{100} \cdot 5b$$

and solve for a:

$$2a + 3b = \frac{13}{10} \cdot 5b$$
$$2a + 3b = \frac{13}{2} \cdot b$$
$$4a + 6b = 13 \cdot b$$
$$4a = 13b - 6b$$
$$4a = 7b$$
$$\frac{a}{b} = \frac{7}{4}$$

Question 14

$AD = 5$ (Pythagorean triple $3 - 4 - 5$), $BC = 5$ (straight segment), and $BD = 5$ (Pythagorean triple $3 - 4 - 5$). The remaining distances are irrational.

Question 15

The area of $\triangle OMC$ is:

$$A_{\triangle OCM} = \frac{OM \cdot CM}{2}$$

and the area of $\triangle OAB$ is:

$$A_{\triangle OAB} = \frac{OB \cdot AM}{2}$$

Since $OB = 2CM$ and $AM = 7$:

$$2\frac{OB \cdot AM}{2} = \frac{OM \cdot CM}{2}$$

$$2 \cdot 2 \cdot OM \cdot 7 = OM \cdot CM$$

$$28 = CM$$

Question 16

Use the definition of the average. The sum of the initial 5 scores is equal to 5 times the average A. After adding a new score, the sum of the scores becomes:

$$5A + 188$$

and the new average is:

$$\frac{5A + 188}{6}$$

Set the condition that the new average is 10% less than the initial average. That is, it is only 90% of the initial average:

$$\frac{5A + 188}{6} = \frac{90}{100}A$$

and solve for A:

$$\frac{5A + 188}{6} = \frac{9}{10}A$$

$$10(5A + 188) = 6 \cdot 9A$$

$$50A + 1880 = 54A$$

$$1880 = 4A$$

$$A = 470$$

Question 17

Although the first term of the sequence can have any parity, the terms following it are *odd, even, odd, even, ...* because multiplication by 2 produces an even number and subtracting 3 from that even produces an odd number.

For the sequence to be periodic we must either have:

an even term is equal to another even term

an odd term is equal to another odd term

since no even number can equal an odd number. Therefore, the period (i.e. number of terms after which a subsequence of terms starts to repeat itself) can only be an even number.

Set the condition that the first and the third terms are equal:

$$1 = 2a - 3$$

and solve for a to find $a = 3$. Indeed, the repeating sequence results:

$$3, \ 6, \ 3, \ 6, \ \cdots$$

You can verify that the same solution is obtained regardless which terms you set equal, provided they are separated by an even number of terms.

Question 18

As the line is translated, it remains parallel to itself, and the larger triangle formed is similar to the original triangle. For the area to increase 9-fold, any linear element of the triangle must increase 3-fold. Therefore, the side OP has to increase from 11 units to 33 units. The line must be translated 22 units to the right.

ALGEBRA QUIZ FIVE

Use the identity $a^2 - b^2 = (a - b)(a + b)$, to transform the following expressions:

1. $1000005^2 - 1000004^2 =$

2. A right angle triangle has a hypotenuse of length 2^7 units and a leg of $2^7 - 1$ units. What is the length of the other leg?

3. $x^4 y^4 - 1 =$

4. $(1005 - 1)(1005 + 1) =$

5. $m^4 - 1 =$

6. $(z^2 + 4)(z^2 - 4) =$

7. $(6 - \sqrt{5})(6 + \sqrt{5}) =$

8. $(\sqrt{11} + \sqrt{10})(\sqrt{11} - \sqrt{10}) =$

9. $(\sqrt[4]{a} - \sqrt[4]{b})(\sqrt[4]{a} + \sqrt[4]{b}) =$

10. $(m^{\frac{3}{4}} - n^{\frac{3}{4}})(m^{\frac{3}{4}} + n^{\frac{3}{4}}) =$

11. 107×93

12. $(m^{\frac{1}{2}} - n)(m^{\frac{1}{2}} + n)$

Answers to questions 1-12

1. 2000009

2. $\sqrt{2^{14} - (2^{14} - 2 \cdot 2^7 + 1)} = \sqrt{2 \cdot 2^7 - 1} = \sqrt{2^8 - 1} =$

    $\sqrt{(2^4 - 1)(2^4 + 1)} = \sqrt{(16 - 5)(16 + 1)} = \sqrt{15 \cdot 17} = \sqrt{155}$

3. $(x^2 y^2 - 1)(x^2 y^2 + 1) = (xy + 1)(xy - 1)(x^2 + y^2)$

4. $1005^2 - 1$

5. $(m^2 + 1)(m^2 - 1) = (m^2 + 1)(m + 1)(m - 1)$

6. $z^4 - 16$

7. $36 - 5 = 31$

8. $11 - 10 = 1$

9. $\sqrt{a} - \sqrt{b}$

10. $m^{\frac{3}{2}} - n^{\frac{3}{2}}$

11. $(100 + 7)(100 - 7) = 10000 - 49 = 9951$

12. $m - n^2$

Geometry Quiz Five

Question 1

The segment that connects the midpoints of two edges of a scalene triangle is .... to the third edge and its length is .... of the length of the third edge.

Question 2

The sum of the interior angles of a convex n-gon (polygon with n sides) is ....

Question 3

The sum of the exterior angles of a convex n-gon is ....

Question 4

Concentric circles have the same ....

Question 5

If two circles are tangent interior, then the distance between their centers is smaller/larger/equal than/to the radius of the larger circle.

Question 6

The opposite angles of a parallelogram are ....

Question 7

The adjacent angles of a parallelogram are ....

Question 8

The reflection of a right angle triangle across a leg forms a rectangle/isosceles triangle/square.

Question 9

The intersection of a cube with the plane that contains two of its opposite edges, is a rectangle/square/rhombus.

Question 10

To produce a sphere, a circle must be rotated around its center/radius/diameter/a point on the circumference.

Question 11

If two identical square pyramids are glued so that their bases coincide, a solid with .... faces and .... edges is formed.

Answers to questions 1-11

Question 1
The segment that connects the midpoints of two edges of a scalene triangle is **parallel** to the third edge and its length is **half** of the length of the third edge.

Question 2
The sum of the interior angles of a convex n-gon (polygon with n sides) is $180° \times (n-2)$.

Question 3
The sum of the exterior angles of a convex n-gon is $360°$.

Question 4
Concentric circles have the same **center**.

Question 5
If two circles are tangent interior, then the distance between their centers is **smaller** than the radius of the larger circle.

Question 6
The opposite angles of a parallelogram are **congruent**.

Question 7
The adjacent angles of a parallelogram are **supplementary**.

Question 8
The reflection of a right angle triangle across a leg is an **isosceles triangle**.

Question 9
The intersection of a cube with the plane that contains two of its opposite edges, is a **rectangle**.

Question 10
To produce a sphere, a circle must be rotated around its **diameter**.

Question 11
If two identical square pyramids are glued so that their bases coincide, a solid is formed. This solid has 8 faces and 12 edges.

DATA ANALYSIS QUIZ FIVE

Combinations are used to *select* k objects out of n objects. The simplest example of combinations is a string of n letters out of which k are of a kind and $n - k$ are of another kind. Example:

NNNNNSSSSSNSNSNNNSS

1. How many letters in total are there in the example?

2. How many of the letter are 'N'?

3. How many of the letter are 'S'?

4. How many possible ways to arrange them in a string are there?

5. If each of the letters 'S' is painted in a different color, how does this change your answer to the question 4. above?

6. In how many different ways can we arrange 3 red teapots, 2 blue teapots, and 4 green teapots, in a row?

7. How many ways are there to arrange 12 toothpicks in a row (all in vertical position)?

8. In how many ways can we arrange 0 objects in a row?

9. Is is true that $0! = 1!$?

10. In how many ways can we select 4 books from a set of 5 identical history books and 3 identical biology books?

Answers to questions

1. 18

2. 10

3. 8

4. The number of combinations is:

$$_{18}C_8 =_{18} C_{10} = \frac{18!}{10!8!}$$

5. Since the 'S' symbols are no longer identical, the number of combinations becomes:

$$_{18}C_8 \cdot 8! = \frac{18!}{10!}$$

6. The total number of objects is $3 + 2 + 4 = 9$. The number of different ways to arrange them in a row is:

$$\frac{9!}{3!2!4!} = 1,260$$

Note that this is not a combination.

7. Just 1, since toothpicks are identical.

8. One way. This is why $0! = 1$.

9. Yes. See the previous question.

10. In 4 ways: BBBH, BBHH, BHHH, HHHH.

Number Sense Quiz Five

If possible, simplify the expressions:

1. $2^{k+1} - 2^{k-1} =$

2. $4^k + 2^{2k} =$

3. $10^{2k} \div 4^k =$

4. $2^{6k} \div 8^k \times 5^{3k} =$

5. $4^m \div \sqrt{2^{2m}} =$

6. $m^0 + 0^m =$

7. $4^{2^m} \times 2^m =$

8. $5^{2p} \times 4^p \div 10^p =$

9. $512^k \div 8^p =$

10. $(k-1)^{m-1} \times (m-1)^{k-1}$

Answers to questions 1-10

1. $2^{k+1} - 2^{k-1} = 2^{k-1}(2^2 - 1) = 3 \cdot 2^{k-1}$

2. Since $4^k = 2^{2k}$, we have $4^k + 2^{2k} = 2 \cdot 2^{2k} = 2^{2k+1}$

3. $10^{2k} \div 4^k = 2^{2k} \times 5^{2k} \div (2^2)^k = 5^{2k}$

4. $2^{6k} \div 8^k \times 5^{3k} = 8^{2k} \div 8^k \times 5^{3k} = 2^{3k} \times 5^{3k} = 10^{3k} = 1000^k$

5. $4^m \div \sqrt{2^{2m}} = 2^{2m} \div 2^m = 2^m$

6. $m^0 + 0^m = 1 + 0 = 1$

7. $4^{2^m} \times 2^m = (2^2)^{2^m} \times 2^m = 2^{2 \times 2^m} \times 2^m = 2^{2^{m+1}+m}$

8. $5^{2p} \times 4^p \div 10^p = 5^{2p} \times 2^{2p} \div 10^p = 10^{2p} \div 10^p = 10^p$

9. $512^k \div 8^p = 2^{9k} \div 2^{3p} = 2^{9k-3p} = 8^{3k-p}$

10. cannot be simplified

DIAGNOSTIC SIX

This math section contains 18 questions to be solved in 25 minutes.

You are allowed to use a calculator. However, in order to minimize the time needed to complete the test, it is important to use the calculator as little as possible! Keep in mind:

- Entering operations in the calculator is more time consuming than performing the operations mentally.

- Data entry errors will be made in addition to other errors. Entering data is, in itself, a possible cause for error.

You can reduce the use of the calculator by memorizing well a short list of commonly used numbers: perfect squares from 1 to 20^2, powers of 2 from 2^0 to 2^{10}, some frequently used Pythagorean triples, etc. The list can be found in Appendix A, as well as on www.mathinee.com.

As you solve problems, build on the ability to recognize the numbers in this list. Every time you encounter them, tell yourself "hey, this is a power of 2", or "hey, this is a Pythagorean triple." After a short while, you will have memorized the list almost completely. But going over the list 'cold' a few times is also helpful, so go ahead and open Appendix A!

Before starting to solve, spend a few seconds to think about focusing, about making sure that you are ready to pay attention to every detail.

1. In a laboratory experiment, 20 rats experienced an increase of 10% in their average lifespan, while 80 rats experienced a 15% increase. What was the average lifespan increase for the 100 rats that participated in the experiment?

(A) 11%

(B) 12%

(C) 12.5%

(D) 13.5%

(E) 14%

2. If $x \cdot y = 26$ what is the value of:

$$\frac{3y^2}{4x^{-4}} \cdot x^{-2}$$

(A) 19.5

(B) 39

(C) 169

(D) 507

(E) $\dfrac{3}{169}$

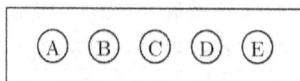

3. If m is 40% of n, which of the following represents 75% of n?

(A) $\dfrac{5}{3}m$

(B) $\dfrac{8}{15}m$

(C) $\dfrac{3}{4}m$

(D) $\dfrac{7}{50}m$

(E) $\dfrac{15}{8}m$

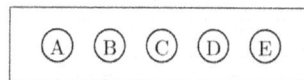

4. Among all the segments that connect two midpoints of edges of a unit cube, how many are there that have a length of $\sqrt{2}$?

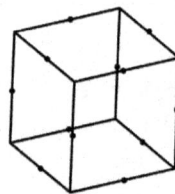

(A) 2 **(B)** 4 **(C)** 6

(D) 8 **(E)** 12

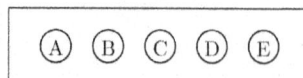

5. An amount q is decreased by 9 and the result is multiplied by 3. 30 is added to the result and the sum is divided by 3. What is the final result, in terms of q?

(A) q

(B) $q + 1$

(C) $q + 3$

(D) $q + 7$

(E) $q + 9$

Ⓐ Ⓑ Ⓒ Ⓓ Ⓔ

6. A small business produces Christmas decorations. Based on the information available from the table below, what was the total number of silver metal ornaments sold during the month of November?

NOVEMBER SALES

	Gold	Silver	Matte	Total
Glass	500	7,200		11,000
Metal	300		8,200	
Plastic	1,200	800	650	
Total				30,500

(A) 6,850

(B) 7,350

(C) 8,000

(D) 8,350

(E) cannot be determined

Ⓐ Ⓑ Ⓒ Ⓓ Ⓔ

7. A square has one vertex on the parabola $y = x^2$ and the vertex digonally opposite to it at $(6,0)$. What is the area of the square?

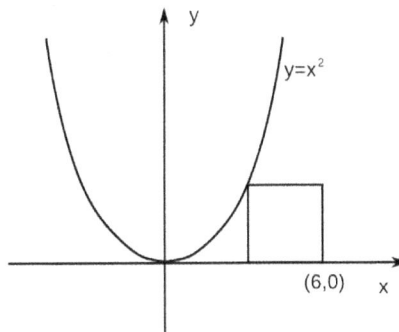

(A) 1 **(B)** 2 **(C)** 4

(D) 16 **(E)** 36

Ⓐ Ⓑ Ⓒ Ⓓ Ⓔ

8. How many integers from 1 to 300 are not the cube of an integer?

(A) 5

(B) 6

(C) 216

(D) 294

(E) 295

Ⓐ Ⓑ Ⓒ Ⓓ Ⓔ

9. l is a line in the plane xOy. The slope of line l is -2. A line m is perpendicular to l and passes through $(2, 0)$. What is the y-intercept of the line m?

(A) -2

(B) -1

(C) $\dfrac{1}{2}$

(D) 1

(E) 2

Ⓐ Ⓑ Ⓒ Ⓓ Ⓔ

10. In an economic depression, house prices decreased at an average rate of 4% per month. If a house used to cost t dollars in September, what would its approximate cost be expected to be in December?

(A) $0.04t$

(B) $0.4t$

(C) $0.66t$

(D) $0.83t$

(E) $0.88t$

Ⓐ Ⓑ Ⓒ Ⓓ Ⓔ

11. If 160 squares form a rectangular row in which any two adjacent squares share one side, how many edges are not shared?

(A) 161

(B) 164

(C) 320

(D) 322

(E) 368

Ⓐ Ⓑ Ⓒ Ⓓ Ⓔ

12. In the Venn diagram below, how many elements belong to A or to B but not to C?

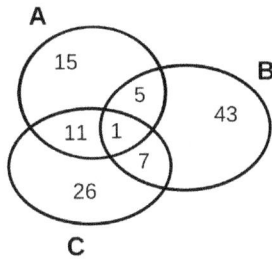

(A) 5 (B) 19 (C) 63
(D) 82 (E) 108

Ⓐ Ⓑ Ⓒ Ⓓ Ⓔ

13. In the isosceles triangle ABC, the segments AM, BN, and CP are perpendicular onto the sides of the triangle. If $AB = AC = 5$ and $BC = 6$, what is the length of BN?

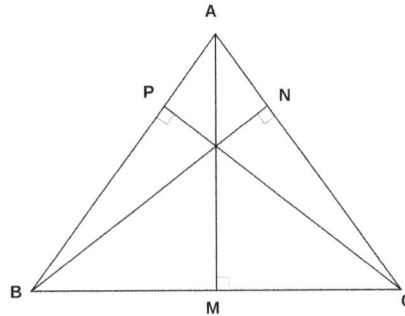

(A) 4
(B) 4.5
(C) 4.8
(D) 5
(E) cannot be determined

Ⓐ Ⓑ Ⓒ Ⓓ Ⓔ

14. A rectangular box has length 4, width 4, and height 9 units. What is the volume of the smallest cube that the box will fit into?

(A) 12
(B) $5\sqrt{10}$
(C) 144
(D) 250
(E) 729

Ⓐ Ⓑ Ⓒ Ⓓ Ⓔ

15. If the figure represents $|f(x)|$, which of the following could be $f(x)$?

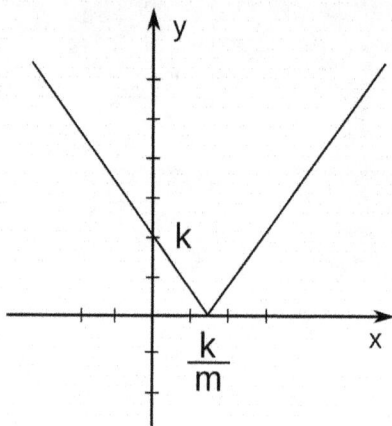

(A) $f(x) = -mx + k$

(B) $f(x) = mx + k$

(C) $f(x) = \dfrac{1}{m}x + k$

(D) $f(x) = -mx - k$

(E) $f(x) = \dfrac{1}{m}x - k$

Ⓐ Ⓑ Ⓒ Ⓓ Ⓔ

16. A vehicle with an average fuel efficiency of 50 mpg in freeway traffic and 40 mpg in city traffic, travels 175 miles on the freeway and 60 miles in the city. What is the average fuel efficiency in mpg for this specific trip?

(A) 45

(B) 47

(C) 48

(D) 49

(E) 50

Ⓐ Ⓑ Ⓒ Ⓓ Ⓔ

29.1 Self-grade Diagnostic Six

Question	1	2	3	4	5	6	7	8
Level	1	1	1	2	2	2	2	3
Topic	N	A	N	G	A	S	A	N
Answer	E	D	E	C	B	D	C	D
Correct								
Incorrect								
Skipped								

Question	9	10	11	12	13	14	15	16
Level	2	3	3	3	4	3	4	4
Topic	A	N	N	S	G	G	A	N
Answer	B	E	D	C	C	E	A	B
Correct								
Incorrect								
Skipped								

Are your incorrect answers:

mostly on questions in the first part of the test?	Yes	No
mostly on questions that have difficulty levels 3 and 4?	Yes	No
mostly from a specific topic (geometry, algebra, arithmetic)?	Yes	No
somewhat evenly spread throughout the test?	Yes	No
mostly due to minor errors?	Yes	No
mostly due to numeric errors?	Yes	No

A-algebra, N-number sense, G-geometry, S-statistics

29.2 Diagnostic Six Solutions

Question 1

Denote the average lifespan with L. 20 rats averaged a $1.1L$ lifespan - the sum of their lifespans totals $20 \times 1.1L = 22L$. 80 rats averaged a $1.15L$ lifespan - the sum of their lifespans totals $80 \times 1.15L = 92L$.

The sum of all lifespans is $22L + 92L = 114L$. The average lifespan is:

$$\frac{114L}{100} = 1.14L$$

which represents an increase of 14% over the normal average lifespan.

Question 2
Simplify the expression:

$$\frac{3y^2}{4x^{-4}} \cdot x^{-2} = \frac{3}{4}y^2 x^{4-2} = \frac{3}{4}x^2 y^2$$

$$= \frac{3}{4}(xy)^2 = \frac{3}{4}(2 \cdot 13)^2 = \frac{3}{4}(4 \cdot 169)$$

$$= 3 \cdot 169 = 507$$

Question 3
Apply the definition of percentages:

$$m = \frac{40}{100}n$$

$$\frac{75}{100}n = \frac{75}{40} \cdot \frac{40}{100}n = \frac{75}{40}m = \frac{15}{8}m$$

Question 4

A segment that connect midpoints of edges has a length of $\sqrt{2}$ if its ends are on diametrically opposite edges.

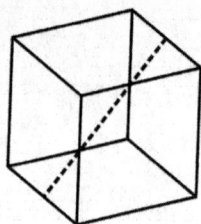

Since there are 12 edges in total, there are 6 pairs of diametrically opposite edges. There is one such segment for each pair, to a total of 6 segments of the required length.

Question 5

Make sure you apply correctly the distributive property:

$$q - 9$$
$$3(q - 9) = 3q - 27$$
$$3q - 27 + 30 = 3q + 3$$
$$3q + 3 = 3(q + 1)$$
$$3(q + 1) \div 3 = q + 1$$

Question 6

The total plastic ornaments: $1200 + 800 + 650 = 2650$. The total metal ornaments: $30500 - 11000 - 2650 = 16850$. The silver metal ornaments: $16850 - 8200 - 300 = 8350$.

Question 7

Denote the side of the square with x. Since the sides of a square have equal lengths:

$$x^2 = 6 - x$$
$$x^2 + x - 6 = 0$$
$$(x - 2)(x + 3) = 0$$

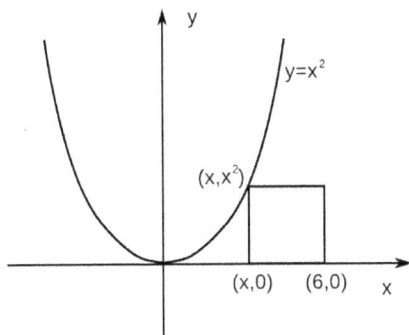

x can be 2 or -3. Since a length cannot be negative, $x = 2$. The area of the square is equal to 4.

Question 8

Cubes increase very fast, therefore we easily find that:

$$6^3 = 216$$

and

$$7^3 = 343$$

Therefore, only 6 numbers between 1 and 300 are perfect cubes. The remaining $300 - 6 = 294$ are not perfect cubes.

Question 9

The equation of the line m in point-slope form is:

$$(y - 0) = \frac{1}{2}(x - 2)$$

We obtain the y-intercept by setting $x = 0$:

$$y = \frac{1}{2}(-2) = -1$$

Question 10

After each month, the price is only 0.96 of the previous month's price. After 3 months, the price would be:

$$0.96 \times 0.96 \times 0.96 \times t \approx 0.885t$$

Question 11

There are 160 unshared edges that form each of two parallel sides of the rectangle and 1 unshared edge that forms each of the other pair of parallel sides. The number of unshared edges is:

$$160 + 160 + 1 + 1 = 322$$

Question 12

There are $15 + 5 + 43 + 11 + 1 + 7 = 82$ elements that belong to A or to B. Of these, $11 + 1 + 7 = 19$ are elements of C. There are $82 - 19 = 63$ elements that belong to A or to B but not to C.

Question 13

The altitude AM is also a median, due to the symmetry of the isosceles triangle. Therefore, $BM = MC = 3$ and the triangles ABM and AMC are $3-4-5$ right angle triangles. Since $AM = 4$, the area of the triangle ABC is:

$$A_{\triangle ABC} = \frac{6 \times 4}{2}$$

The same area can also be written as:

$$A_{\triangle ABC} = \frac{5 \times BN}{2}$$

Solve for BN:
$$6 \times 4 = 5 \times BN$$

$BN = 4.8$ units.

Question 14

The cube has to be sufficiently large to fit the side of length 9 units.
The volume of the cube is at least $9 \times 9 \times 9 = 729$ cubic units.

Question 15

Since the x-intercept is $\dfrac{k}{m}$:

$$
\begin{aligned}
mx_{\text{int}} &= k \\
mx_{\text{int}} - k &= 0
\end{aligned}
$$

or

$$-mx_{\text{int}} + k = 0$$

The two lines that can have the given x-intercept are:

$$
\begin{aligned}
f(x) &= mx - k \\
f(x) &= -mx + k
\end{aligned}
$$

Of these, only the second one is an answer choice.

The graph may be the result of taking the absolute value of either of the two functions:

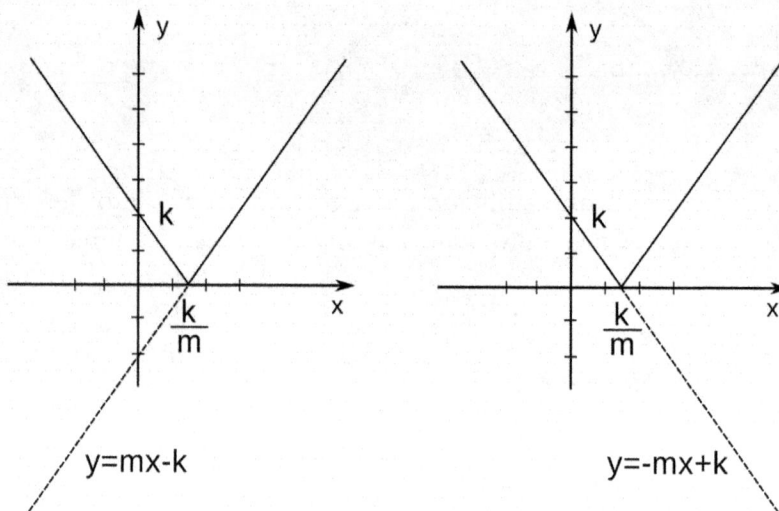

Question 16

The fuel consumption, in gallons, is obtained by dividing the mileage by the mpg. The fuel consumption on the freeway was:

$$\frac{175}{50}$$

and the fuel consumption in the city was:

$$\frac{60}{40}$$

The total amount of fuel used was:

$$\frac{175}{50} + \frac{60}{40} = \frac{35}{10} + \frac{15}{10} = \frac{50}{10} = 5 \text{ gal}$$

The total mileage was:

$$175 + 60 = 235 \text{ miles}$$

The average mpg was:

$$\frac{235}{5} = 47 \text{ mpg}$$

Algebra Quiz Six

1. If a line has negative slope, which quadrants of the coordinate plane does it *necessarily* cross?

2. If a line passes through the origin of the coordinate system, then its y-intercept and its x-intercept ....

3. A parabola and a line that are in the same plane intersect no more than .... times.

4. Two parabolas may intersect at: 0/1/2/3/4/5/6 points (circle all that apply.)

5. If two lines have opposite slopes (such as a and $-a$), they form a figure with 0/1/2/3 axes of symmetry (circle all that apply.)

6. Two lines pass through the origin and have slopes that are reciprocals of one another. The two lines may pass through quadrants:

   (A) I and II
   (B) II and IV
   (C) IV and III
   (D) all four quadrants

7. If the line $y = 7$ intersects a line parallel to the y-axis that is 4 units away from the y-axis, what are the coordinates of the possible intersection points?

8. A function cannot have:

   (A) the same value for two different values of the argument x
   (B) two different values for the same value of the argument x
   (C) one value that is larger than all its other values
   (D) only negative arguments

Answers to questions 1-8

1. Quadrants II and IV.

2. If a line passes through the origin of the coordinate system, then its y-intercept and its x-intercept **coincide** (that is, they are the same point.)

3. A parabola and a line that are in the same plane intersect no more than 2 times.

4. $0, 1, 2$

5. 2

6. If the slopes are reciprocals then they are either both positive or both negative. The lines may pass through quadrants I and III or II and IV. Of these, only one is an answer choice.

7. The intersection points could be either $(4, 7)$ or $(-4, 7)$, depending on whether the vertical line is on the right hand side of the y-axis or on the left side, respectively.

8. A function cannot have two different values for the same value of the argument x. This is also known as 'the vertical line test.' All other choices are valid possibilities.

Geometry Quiz Six

1. The interior angle of a regular polygon with 9 sides is . . . .

2. A triangle has at most . . . obtuse angle(s).

3. What is the number of axes of symmetry of a right circular cone?

4. If we cut off a single vertex of a cube with a plane, how many edges does the resulting solid have?

5. Two circles that have the same center and the same radius as a sphere are perpendicular to one another. In how many parts do the circles divide the sphere?

6. How many distinct triangles can be formed using 5 points that are in the same plane but no 3 of them are on the same line?

7. The set of points that are equidistant from a given line is a line/two lines/a point/a circle.

8. The set of points that are equidistant from a circle is a circle/two circles/three circles/a circle and a point/a point.

Answers to questions 1-8

1. 140°

2. 1

3. One axis: the line connecting the vertex of the cone and the center of the base circle.

4. The cube has 12 edges. By cutting off a vertex, three new edges are formed. The total number of edges is now 15.

5. 4 parts. Do this with an apple to visualize the process.

6. $_5C_3 = \dfrac{5!}{3!2!} = 10$

7. The set of points that are equidistant from a given line is formed of two lines that are parallel to the given line, one on each side of it.

8. The set of points that are equidistant from a circle is formed of two circles that are concentric with the given circle, one inside and one outside.

NUMBER SENSE QUIZ SIX

1. Find the largest power of 8 that is a factor of 60^{12}.

2. Find the largest power of 3 that is a factor of 405^6.

3. Find the largest power of 4 that is a factor of 72^{16}.

4. What is the largest common factor of $216, 360$, and 900?

5. What is the least common multiple of $49, 21$, and 111?

6. What is the lcm of the first 5 positive multiples of 3?

7. What is the product of all the multiples of 5 between -19 and 19?

8. How many positive factors/divisors does the number 28 have?

9. If $\tau(N)$ represents the number of positive factors/divisors of N, what is $\tau(\tau(15))$?

10. How many factors/divisors does the number 16 have?

Answers to questions 1-10

1. Factor 60 into primes and apply the properties of exponents:
$$60^{12} = (2^2 \cdot 3 \cdot 5)^{12} = 2^{24} \cdot 3^{12} \cdot 5^{12}$$
Since: $2^{24} = (2^3)^8 = 8^8$, the largest power of 8 that divides 60^{12} is 8^8.

2. 24 since: $405^6 = 3^{24} \cdot 5^6$.

3. 24 since $(2^3 \cdot 3^2)^{16} = 2^{48} \cdot 3^{32} = 4^{24} \cdot 3^{32}$.

4. $\gcf(216, 360, 900) = \gcf(2^3 \cdot 3^3, 2^3 \cdot 3^2 \cdot 5, 2^2 \cdot 3^2 \cdot 5^2) = 2^2 \cdot 3^2 = 36$

5. $\lcm(49, 21, 111) = \lcm(7^2, 7 \cdot 3, 3 \cdot 37) = 7^2 \cdot 3 \cdot 37 = 5439$

6. $\lcm(3, 6, 9, 12, 15) = \lcm(3, 2 \cdot 3, 3^2, 2^2 \cdot 3, 3 \cdot 5) = 2^2 \cdot 3^2 \cdot 5 = 180$

7. 0, since 0 is a multiple of 5 within the specified range.

8. Since $28 = 2^2 \cdot 7$, it has 6 positive divisors: $\{1, 2, 4, 7, 14, 28\}$.

9. $\tau(\tau(15)) = \tau(4) = 3$

10. There are 10 divisors: $\{-16, -8, -4, -2, -1, 1, 2, 4, 8, 16\}$.

DATA ANALYSIS QUIZ SIX

Answer the questions for each of the Venn diagrams:

There are 15 elements in A, 29 elements in B, and 32 elements in the union A∪B. How many elements are there in the intersection of A and B?

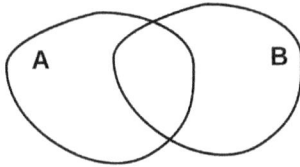

There are 51 elements in C, 92 elements in D, and 32 elements in the intersection of C and D. How many elements are there in the union of C and D?

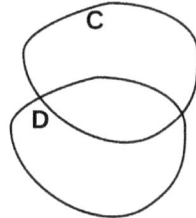

How many elements belong to either F or H but not to G?

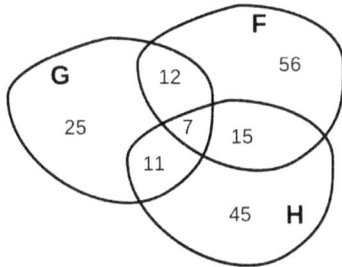

How many elements belong to Q or to W and to T?

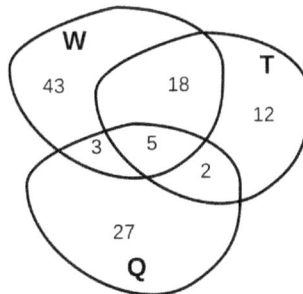

L has 45 elements and K has 81 elements. At most, how many elements are in the intersection L∩K?

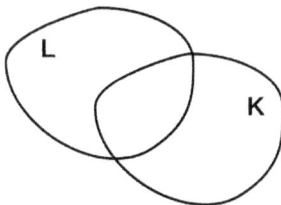

How many elements belong to S or to P, but not to both?

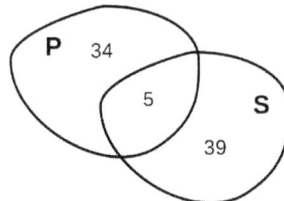

Answers to questions

The intersection A ∩ B has:
15+29-32=12 elements.

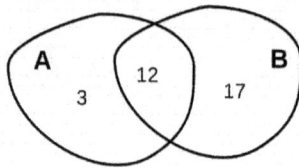

The union AUB has:
19+32+60=111 elements.

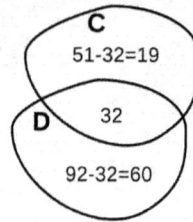

56+15+45=116 elements
belong to either F or H
but not to G.

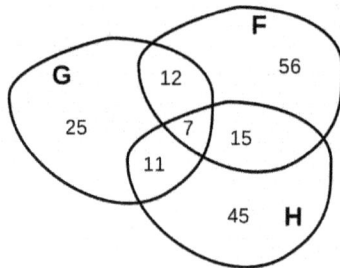

27+3+5+2+18=55 elements
belong to Q or to W and to T.

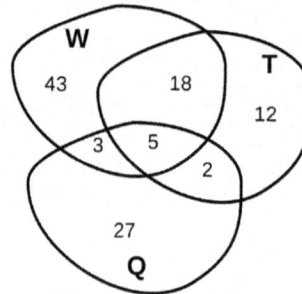

There are at most 45 elements
in the intersection L∩K.

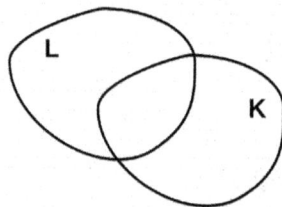

34+39=73 elements belong
to S or to P but not to both.

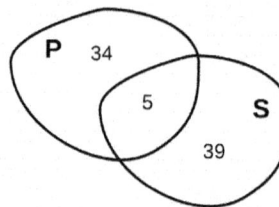

DIAGNOSTIC SEVEN

This math section contains 18 questions to be solved in 25 minutes.

You are allowed to use a calculator. However, in order to minimize the time needed to complete the test, it is important to use the calculator as little as possible! Keep in mind:

- Entering operations in the calculator is more time consuming than performing the operations mentally.

- Data entry errors will be made in addition to other errors. Entering data is, in itself, a possible cause for error.

You can reduce the use of the calculator by memorizing well a short list of commonly used numbers: perfect squares from 1 to 20^2, powers of 2 from 2^0 to 2^{10}, some frequently used Pythagorean triples, etc. The list can be found in Appendix A, as well as on www.mathinee.com.

As you solve problems, build on the ability to recognize the numbers in this list. Every time you encounter them, tell yourself "hey, this is a power of 2", or "hey, this is a Pythagorean triple." After a short while, you will have memorized the list almost completely. But going over the list 'cold' a few times is also helpful, so go ahead and open Appendix A!

Before starting to solve, spend a few seconds to think about focusing, about making sure that you are ready to pay attention to every detail.

1. If $x = -4$ and $y = 0.25$ what is the value of:

$$x^{66} \cdot y^{66}$$

(A) -512

(B) $-\dfrac{1}{8}$

(C) -1

(D) 1

(E) 1024

Ⓐ Ⓑ Ⓒ Ⓓ Ⓔ

2. There are 18 people who volunteered to repaint the house numbers on the sidewalk and 10 people who volunteered to check the street water drains. At least how many volunteers should change their option in order for both teams to have the same number of workers?

(A) 2

(B) 4

(C) 5

(D) 6

(E) 8

Ⓐ Ⓑ Ⓒ Ⓓ Ⓔ

3. If $5^{x+3} = 625$ and $(-x)^5 = z$, what is the value of z^{-x}?

(A) -5

(B) -1

(C) $\dfrac{1}{5}$

(D) 1

(E) 5

Ⓐ Ⓑ Ⓒ Ⓓ Ⓔ

4. Which of the following graphs could represent a function that is even and has period 4 units?

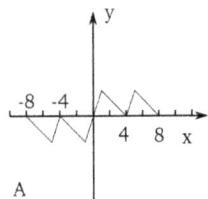

A

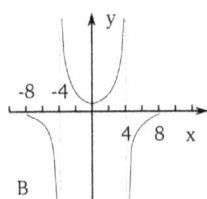

B

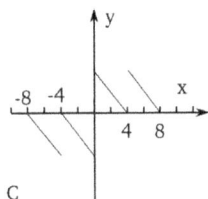

C

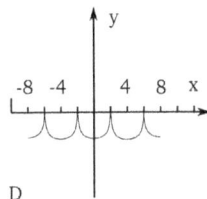

D

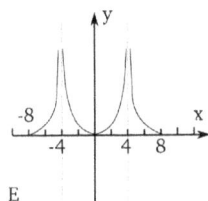

E

(A) A

(B) B

(C) C

(D) D

(E) E

Ⓐ Ⓑ Ⓒ Ⓓ Ⓔ

5. In a field, there are brown, white, black and grey cows. There are one more brown cows than cows of any other color. If 10 cows have the same color, what is the minimum number of cows in the field?

(A) 27

(B) 30

(C) 34

(D) 37

(E) 41

Ⓐ Ⓑ Ⓒ Ⓓ Ⓔ

6. Four riders have to choose among six horses. How many possible choices are there?

(A) 15

(B) 24

(C) 360

(D) 576

(E) 720

Ⓐ Ⓑ Ⓒ Ⓓ Ⓔ

7. If an equilateral triangle rotates clockwise around its center 15° at a time, how many rotations are necessary for it to be back to its initial position?

(A) 6

(B) 8

(C) 12

(D) 24

(E) 180

Ⓐ Ⓑ Ⓒ Ⓓ Ⓔ

8. Compute the value of

$$\sum_{k=-4}^{4} (k^2 - 1)$$

(A) −9 **(B)** 0 **(C)** 1
(D) 51 **(E)** 52

Ⓐ Ⓑ Ⓒ Ⓓ Ⓔ

9. A rectangle has sides of lengths x and y. If both lengths are doubled, then the resulting area is $p\%$ larger than the initial area. What is p?

(A) 4

(B) 100

(C) 200

(D) 300

(E) 400

Ⓐ Ⓑ Ⓒ Ⓓ Ⓔ

10. In 3D space, the set of points that are equidistant from three distinct, non-collinear points is:

(A) a point

(B) a line

(C) a circle

(D) a sphere

(E) a plane

Ⓐ Ⓑ Ⓒ Ⓓ Ⓔ

11. If $f(x) = \sqrt{x+4}$ and $g(x) = x - 5$, which of the following operations may not be possible?

(A) $f \circ g$

(B) $g - 4$

(C) $\dfrac{g}{f+1}$

(D) $g \circ f$

(E) $g^{-1} \circ f$

Ⓐ Ⓑ Ⓒ Ⓓ Ⓔ

---

12. The table below summarizes the ratio of success among the teams that attempted to climb the Eiger peak over the last 5 decades.

In which decade was the percentage of unsuccessful expeditions equal to 80%?

decade	successful attempts
2000-2010	three fifths
1990-2000	two fifths
1980-1990	four fifths
1970-1980	one fifth
1960-1970	five eights

(A) 2000-2010

(B) 1990-2000

(C) 1980-1990

(D) 1970-1980

(E) 1960-1970

Ⓐ Ⓑ Ⓒ Ⓓ Ⓔ

13. The terms of a sequence are generated by alternately adding 5 and subtracting 2. The sequence starts at 1 and continues as follows:

$$1, 6, 4, 9, 7, \cdots$$

The 99th term of this sequence is:

(A) 148   **(B)** 150   **(C)** 297

(D) 303   **(E)** 385

Ⓐ Ⓑ Ⓒ Ⓓ Ⓔ

---

14. Water represents 99% of the weight of a full sponge. The wet sponge weighs 600 grams. Some of the water evaporates until only 98% of the wet sponge's weight is due to the water content. What is the weight of the sponge after the evaporation?

(A) 300 g

(B) 400 g

(C) 500 g

(D) 594 g

(E) 598.06 g

Ⓐ Ⓑ Ⓒ Ⓓ Ⓔ

15. The average speed of a vehicle over a 600 mile journey is 100 mph. If its average speed over half of the route was 75 mph, what was its average speed, in mph, over the remaining distance?

(A) 110

(B) 125

(C) 133

(D) 140

(E) 150

$$\text{(A)} \quad \text{(B)} \quad \text{(C)} \quad \text{(D)} \quad \text{(E)}$$

16. Line l has a y-intercept of 5 and line p has a y-intercept of -2. The two lines intersect at $(3, 4)$. What is the product of their slopes?

(A) $-\dfrac{2}{3}$

(B) $\dfrac{2}{3}$

(C) $-\dfrac{1}{4}$

(D) $\dfrac{5}{16}$

(E) cannot be determined

$$\text{(A)} \quad \text{(B)} \quad \text{(C)} \quad \text{(D)} \quad \text{(E)}$$

17. What is the distance between the lines $y = x$ and $y = x + 5$?

(A) -5

(B) 1

(C) 2.5

(D) $\dfrac{5}{\sqrt{2}}$

(E) 5

$$\text{(A)} \quad \text{(B)} \quad \text{(C)} \quad \text{(D)} \quad \text{(E)}$$

18. A toddler plays with magnets. He makes a chain out of magnets. Then he makes another chain that has double the number of magnets of the first chain. Which of the following could be the number of magnets he used in total?

(A) 40

(B) 93

(C) 107

(D) 113

(E) cannot be determined

$$\text{(A)} \quad \text{(B)} \quad \text{(C)} \quad \text{(D)} \quad \text{(E)}$$

19. The diagonal of a rectangle of area 16 square units is the side of a square. The area of the square is:

(A) always larger than 32

(B) always smaller than 16

(C) always smaller than 8

(D) always equal to 16

(E) always equal to 32

20. For the target in the figure, the probability of hitting the shaded area is p. If the side of the square is increased 3 times the probability of hitting the same shaded area is pk. What is the value of k? (The probability of hitting any one point within the square is the same.)

(A) cannot be determined

(B) $\dfrac{1}{9}$

(C) $\dfrac{1}{3}$

(D) 3

(E) 9

34.1   Self-grade Diagnostic Seven

Question	1	2	3	4	5	6	7	8	9	10	11	12
Level	1	1	1	2	1	2	2	2	2	3	3	2
Topic	N	N	A	A	N	S	G	A	N	G	A	S
Answer	D	B	B	D	D	C	B	D	D	B	A	D
Correct												
Incorrect												
Skipped												

Question	13	14	15	16	17	18	19	20
Level	3	4	3	3	4	3	4	4
Topic	N	N	N	A	A	N	G	S
Answer	A	A	E	A	D	B	A	B
Correct								
Incorrect								
Skipped								

Are your incorrect answers:

mostly on questions in the first part of the test?	Yes	No
mostly on questions that have difficulty levels 3 and 4?	Yes	No
mostly from a specific topic (geometry, algebra, arithmetic)?	Yes	No
somewhat evenly spread throughout the test?	Yes	No
mostly due to minor errors?	Yes	No
mostly due to numeric errors?	Yes	No

A-algebra, N-number sense, G-geometry, S-statistics

34.2 Diagnostic Seven Solutions

Question 1

Since:

$$x^{66} \cdot y^{66} = (x \cdot y)^{66}$$

We have:

$$(-4 \cdot 0.25)^{66} = (-1)^{66} = 1$$

Question 2

The total number of people is 28. If we want to make two equal groups, there must be 14 people in each group. 4 people must move from the more numerous group to the smaller group.

Question 3

Since

$$
\begin{aligned}
5^4 &= 625 \\
x &= 1 \\
z &= (-x)^5 = (-1)^5 = -1 \\
z^{-x} &= (-1)^{-1} = \frac{1}{-1} = -1
\end{aligned}
$$

Question 4

An even function has the same value for x as for $-x$. Therefore, it is symmetric with respect to the y-axis. The functions that are symmetric with respect to the y-axis are (B), (D), and (E). Of these, the only one that has a period of 4 is (D). (E) appears to have a period of 8, while it is not clear whether (B) is periodic.

Question 5

The smallest number of cows is obtained when the brown cows are the most numerous and all the other cows are one cow less than the brown. Make 10 the number of brown cows and there will be 9 white, 9 black, and 9 grey cows, to a total of $27 + 10 = 37$ cows.

Question 6

Select 4 out of 6 horses:

$$_6C_4 = \frac{6!}{4!2!}$$

and multiply by the number of possible permutations of the riders (of the selected horses, the first rider has 4 to pick from, the second rider has 3 to pick from, etc.):

$$\frac{6!}{4!2!} \times 4! = \frac{6!}{2!} = 360$$

Question 7

An equilateral triangle must rotate around its center in the same direction for $120°$ to be back to its initial position. This requires 8 rotations of $15°$ each.

Question 8

Give values from -4 to 4 to k to find each term of the sum:

$$\sum_{k=-4}^{4} (k^2 - 1) = (-4)^2 - 1 + (-3)^2 - 1 + \cdots + 3^2 - 1 + 4^2 - 1$$

There are 9 terms in total in the sum. The squared terms are all positive. We have:

$$2 \times (1 + 4 + 9 + 16) - 9 = 2 \times 30 - 9 = 60 - 9 = 51$$

Question 9

The initial area of the rectangle is xy. After doubling the side lengths, the area becomes $2x \times 2y = 4xy$, i.e. it is 4 times larger than the initial area. This means it is 300% larger.

Question 10

In a plane, there is a single point that is equidistant from three distinct non-collinear points: the center of the circumcircle of the triangle formed by the points.

In space, the points of a line that passes through the circumcircle and is perpendicular to the triangle are all equidistant from the vertices of the triangle:

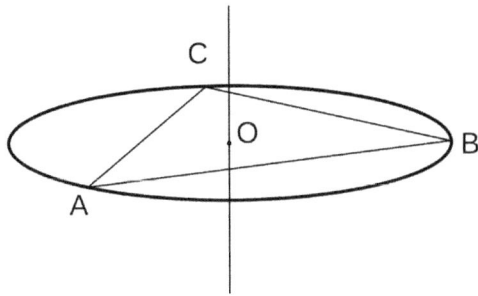

In the figure, A, B, and C are the given points, O is the center of the circumcircle of $\triangle ABC$, and the line passing through O and perpendicular to the circle is the required locus.

Question 11

Since f is defined only for numbers that satisfy $x \geq -4$, it is not possible to apply f to g. The range of g is all real numbers. Thereofore, the choice $f \circ g$ is not possible.

It is, however possible to apply g to f, or g^{-1} to f, since both g and its inverse are defined for all real numbers.

Question 12

Since:
$$\frac{80}{100} = \frac{4}{5}$$

we must look for the decade with $\frac{1}{5}$ successful attempts. This is 1970-1980.

Question 13

Notice that the terms of even rank form a sequence of multiples of 3 starting at 6:

$$6, 9, 12, \cdots$$

The 100th term is part of this subsequence and it is equal to:

$$6 + 49 \times 3 = 6 + 147 = 153$$

The 99th term of the sequence is 5 less than the 100th term:

$$153 - 5 = 148$$

Question 14

Initially, the mass of the water was:

$$600 \times \frac{99}{100} = 6 \times 99 = 594$$

and the mass of the dry sponge is $600 - 594 = 6$ grams. Denote the new mass of the wet sponge after evaporation by $x + 6$, where x is the new amount of water. If the water is 98% of the total mass, then:

$$x = \frac{98}{100}(x + 6)$$

Solve for x:

$$
\begin{aligned}
100x &= 98x + 98 \times 6 \\
2x &= 98 \times 6 \\
x &= 98 \times 3 = 294 \text{ grams}
\end{aligned}
$$

The weight of the sponge after evaporation is $294 + 6 = 300$ grams.

Question 15

The total time needed to travel was:

$$600 \div 100 = 6 \text{ hours}$$

If the vehicle averaged 75 mph over 300 miles, then the time needed to travel these 300 miles was:

$$300 \div 75 = 4 \text{ hours}$$

This means the remaining 300 miles were traveled in 2 hours. The average speed over this distance was 150 mph.

Question 16

The slopes of l and p are, respectively:

$$\frac{4-5}{3} = -\frac{1}{3}$$

$$\frac{4-(-2)}{3} = 2$$

Their product is:

$$\frac{-1}{3} \times 2 = -\frac{2}{3}$$

Question 17

The lines have slope 1 and are parallel:

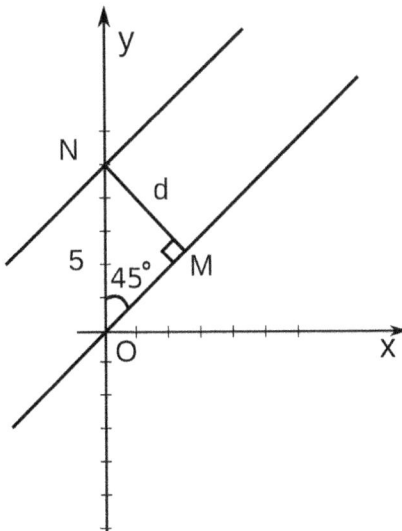

The distance between two parallel lines is the length of the segment which is perpendicular to both. In the figure, it is the segment denoted with d. Since the slope is 1, both lines make an angle of 45° with the y-axis, and the triangle ΔOMN is $45° - 45° - 90°$. d is equal to $\dfrac{5}{\sqrt{2}}$.

Question 18

If the second chain has double the number of magnets of the first chain, then the total number of magnets must be a multiple of 3:

$$x + 2x = 3x$$

The only multiple of 3 among the answer choices is 93. Quickly add the number of digits of each choice to find out which one is divisible by 3.

Question 19

Assume the sides of the rectangle have lengths a and b. Then the diagonal is $\sqrt{a^2 + b^2}$ units long and the area of the square with a side equal to the diagonal is $a^2 + b^2$. Since:

$$(a - b)^2 \geq 0$$
$$a^2 - 2ab + b^2 \geq 0$$
$$a^2 + b^2 \geq 2ab$$

the area of the square is always larger than 32. The equality would happen if and only if both the rectangle and the square would have zero sides.

Question 20

The geometric probability p is tha ratio of the shaded area and the area of the square. If the length of the side of the square triples, the area of the square becomes 9 times larger and the probability decreases 9 times. The factor k is equal to $\dfrac{1}{9}$.

ALGEBRA QUIZ SEVEN

If $f(g(x)) = x$ then f and g are the inverses of one another. Find the expression of the function f in each of the following functional equations:

1. $g(f(x)) = x$ and $g(x) = -x + 5$

2. $f(g^{-1}(x)) = -x$ and $g(x) = 2x - 2$

3. $f(g(x)) = \dfrac{x}{2}$ and $g(x) = 2x$

4. $g(f^{-1}(x)) = x - 1$ and $g(x) = 3 - x$

5. $f(g(x)) = x$ and $g(x) = -x$

If $f(a) = xa^2 + ya + z$, which of the following statements is true for any real value of a?

6. If $x < 0$ there is a value of f that is the largest possible for any value of a.

7. If $z = 0$, the graph of $f(a)$ is tangent to the a-axis.

8. If $x < 0$ and $y^2 - 4xz = 0$, the largest possible value of f is zero.

Answers to questions 1-8

1. $-f(x) + 5 = x$, $f(x) = -x + 5$. Note that f is its own inverse.

2. $g = 2x - 2$, $x = \dfrac{g + 2}{2}$, $g^{-1}(x) = \dfrac{x + 2}{2}$, $f\left(\dfrac{x + 2}{2}\right) = -x$, $f\left(\dfrac{x + 2}{2}\right) = -2\left(\dfrac{x + 2}{2}\right) + 2$, $f(x) = -2x + 2$

3. $f(2x) = \dfrac{x}{2}$, $f(2x) = \dfrac{2x}{4}$, $f(x) = \dfrac{x}{4}$

4. $g(f^{-1}(x)) = x - 1 = 3 - f^{-1}(x)$, $f^{-1}(x) = 4 - x$, $x = f(4 - x)$, $f(4 - x) = x - 4 + 4 = -(4 - x) + 4$, $f(x) = -x + 4$

5. $f(-x) = x$, $f(x) = -x$

6. True. The parabola is convex and its vertex is a maximum.

7. False. Graph, for example, $a^2 - a$.

8. True. The discriminant is zero, the parabola is tangent to the a-axis at its vertex.

Geometry Quiz Seven

Statement	True	False	Sometimes true
Two circles traced on a sphere are congruent.			
The measure of an angle formed by a tangent and a chord equals half the measure of the intercepted arc.			
A rhombus is a parallelogram.			
A kite is a parallelogram.			
The diagonals of a kite are perpendicular.			
The diagonals of a parallelogram bisect each other.			
Any two opposite sides of a parallelogram are congruent.			
Any two adjacent sides of a rectangle are congruent.			
Any two equilateral triangles are similar.			
Two skew lines never intersect.			

Answers to questions

Statement	True	False	Sometimes true
Two circles traced on a sphere are congruent.			V
The measure of an angle formed by a tangent to a circle and a chord of the same circle equals half the measure of the intercepted arc.	V		
A rhombus is a parallelogram.	V		
A kite is a parallelogram.			V (if it is a rhombus)
The diagonals of a kite are perpendicular.	V		
The diagonals of a parallelogram bisect each other.	V		
Any two opposite sides of a parallelogram are congruent.	V		
Any two adjacent sides of a rectangle are congruent.			V (if it is a square)
Any two equilateral triangles are similar.	V		
Two skew lines never intersect.	V		

Number Sense Quiz Seven

1. What is the sum of the smallest prime, the smallest perfect square, the smallest positive integer, and the smallest composite number?

2. What is the number that has a quotient of 11 and a remainder of 7 when divided by 9?

3. What is the number that has a quotient of 9 and a remainder of 7 when divided by 11?

4. What is the number that has a quotient of 4 and a remainder of 8 when divided by 10?

5. If a number has a quotient of 14 and a remainder of 3 when divided by 5, what is the largest prime the number is divisible by?

6. How many different numbers have a quotient of 13 when divided by 7?

7. What is the largest power of 3 that divides 9!?

8. If $\dfrac{9!}{k!} = 504$ what is the value of k?

9. For how many integers k is the fraction $\dfrac{6}{k+1}$ an integer?

10. 11 consecutive numbers have an average of 66. What is the middle number?

Answers to questions 1-10

1. $2 + 0 + 1 + 4 = 7$

2. $9 \times 11 + 7 = 99 + 7 = 106$

3. $9 \times 11 + 7 = 99 + 7 = 106$

4. $4 \times 10 + 8 = 48$

5. The number is $14 \times 5 + 2 = 73$. Since 73 is prime, the largest prime it is divisible by is 73.

6. 7; one for each possible remainder.

7. 4; there is a factor of 3 in each of 3 and 6 and two factors of 3 in 9.

8. 6 because $504 = 7 \times 8 \times 9$ and:

$$\frac{9!}{6!} = \frac{\cancel{6!} \cdot 7 \cdot 8 \cdot 9}{\cancel{6!}} = 7 \cdot 8 \cdot 9 = 504$$

9. 8 values: $-7, -4, -3, -2, 0, 1, 2, 5$

10. 66. If the number of numbers is odd (in our case it is 11), the sum of the numbers is equal to the middle number multiplied by the number of numbers. Therefore, the average is equal to the middle number.

DATA ANALYSIS QUIZ SEVEN

1. If there are 5 Tuesdays in a month of March, which of the following cannot be a weekday that also occurs 5 times in that month?

(A) Sunday

(B) Monday

(C) Wednesday

(D) Thursday

(E) Friday

2. If Plautus wants to paint the three windows of his house in different colors and he has 6 colors to choose from, how many different choices can he make?

3. There are 5 different trails going from White Peak to Desolation Rock and 3 different trails from Desolation Rock to Frog Gulch. How many different trails are there going from White Peak to Frog Gulch through Desolation Rock?

4. There are 50 red chips and 50 black chips in a bag. If we remove chips blindfolded, at least how many chips do we have to remove to be sure we have at least 3 chips of each color?

5. There are 50 red chips and 50 black chips in a bag. After removing 12 chips blindfolded, which of the following outcomes is not possible:

(A) to remove 2 times as many red chips as black chips

(B) to remove 2 times as many black chips as red chips

(C) to remove 3 times as many red chips as black chips

(D) to remove 3 times as many black chips as red chips

(E) to remove 6 times as many red chips as black chips

Answers to questions 1-5

1. Friday. Make a calendar that may have Tuesdays as the first, second, or third day of March.

2. He has 6 choices for the first window, 5 choices for the second one, and 4 choices for the third, for a total of $6 \times 5 \times 4 = 120$ choices.

We can also use $_6C_3 = 20$ and multiply by $3! = 6$ which is the number of possible permutations of the chosen colors when applying them to specific windows.

3. 15

4. 53

5. It is impossible to remove 6 times as many red chips as black chips since the total number of removed chips would have to be a multiple of 7.

DIAGNOSTIC EIGHT

This math section contains 18 questions to be solved in 25 minutes.

You are allowed to use a calculator.

The first 8 questions are multiple choice, the remaining 10 questions are student response questions. For the student response questions, the answer consists of 4 characters chosen from the symbols: . /, and any of the ten digits. The answer may not start with a zero. Answers that have both a fractional and a decimal representation may be represented either way. For example, the representations:

$$1/4$$

and

$$.25$$

are equivalent. Note that 0.25 is not a possible entry.

Before starting to solve, spend a few seconds to think about focusing, about making sure that you are ready to pay attention to every detail.

1. If $a = -3$ and $b = 2$ which of the following is a^{-1}/b^{-1}?

(A) $-\dfrac{3}{2}$

(B) $-\dfrac{2}{3}$

(C) $-\dfrac{1}{6}$

(D) $\dfrac{2}{3}$

(E) $\dfrac{3}{2}$

Ⓐ Ⓑ Ⓒ Ⓓ Ⓔ

2. If ∇ is an operator such that:

$$m\nabla n = n^2 - m^2$$

Then what is $4\nabla 2\nabla 1$?

(A) -143

(B) -25

(C) -7

(D) 11

(E) 143

Ⓐ Ⓑ Ⓒ Ⓓ Ⓔ

3. How many numbers can there NOT be in a sequence of consecutive numbers if there are 61 multiples of 3 among them?

(A) 184

(B) 185

(C) 186

(D) 187

(E) 188

Ⓐ Ⓑ Ⓒ Ⓓ Ⓔ

4. In the xy-plane, line l is perpendicular to the line $y = ax + t$ and parallel to the line $x = by + m$. Which of the following must be true?

(A) $a = -b$

(B) $a = -\dfrac{1}{b}$

(C) $b = \dfrac{1}{a}$

(D) $a = \dfrac{m}{b}$

(E) $b = -\dfrac{t}{ma}$

Ⓐ Ⓑ Ⓒ Ⓓ Ⓔ

5. The average grade of a student is 85% over 6 tests. On his last test, he obtained a grade of 90%. What was his average grade before the last test?

(A) 78%

(B) 79%

(C) 80%

(D) 82%

(E) 84%

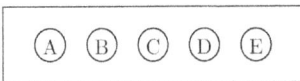

6. If $f(x) = \frac{x}{|x|}$ for any $x \neq 0$, which of the following may not be true for some value of x:

(A) $f(-\frac{1}{3}) = -1$

(B) $f((-10)^5) = -1$

(C) $f(4) \cdot (f(-4) = f(10) \cdot f(-10)$

(D) $f(x + 1) = f(x - 1)$

(E) $f(x^2) = f(\frac{1}{x^2})$

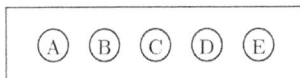

7. If a car drives through 4 successive sets of lights that can only display red, green or amber, how many possible combinations of different light settings can it encounter?

(A) 3

(B) 7

(C) 12

(D) 18

(E) 81

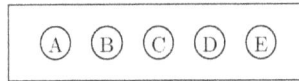

8. In the figure below $\triangle ABC$ is an equilateral triangle with side length $\sqrt{3}$, M is the midpoint of AB, MN is parallel to BC, and the segments MN and BC are congruent.

What is the area of the parallelogram $BCNM$?

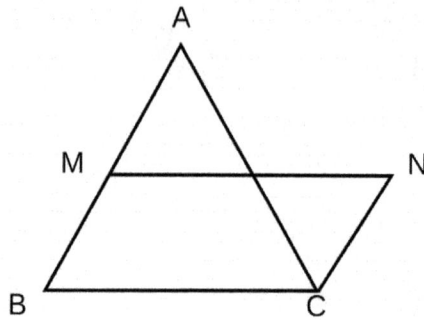

(A) $\dfrac{3}{2}$

(B) $\dfrac{3\sqrt{3}}{2}$

(C) $\dfrac{3}{4}$

(D) $\dfrac{3\sqrt{3}}{4}$

(E) 3

Ⓐ Ⓑ Ⓒ Ⓓ Ⓔ

9. Three radii divide a circle in three parts as in the figure (not drawn to scale). Arc a is longer than arc b by 35% while arc c is shorter than arc b by 35%. How many degrees of arc is the measure of the angle b?

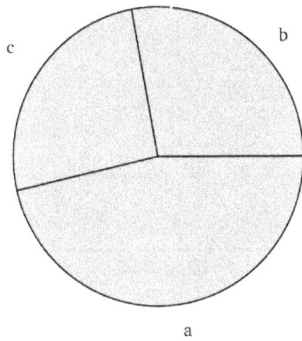

10. How many even integer numbers satisfy

$$|x - 2| < 7$$

11.   Points A, B, and D are on the line (l) such that the length of the segment AB is 2.5 times as large as the length of the segment BD. Point P is twice as far from the line (l) as point Q. By how much percent is the area of $\triangle ABP$ larger than the area of $\triangle BDQ$?

12.   A bakery baked 12% more loaves than ordered. 5% of the total number were rejected during weighing. In percents, how many more loaves than the ordered number does the bakery have after eliminating the rejected loaves?

13.   If $|x - 3| < 4$, then $|x + 7| < A$. What is the smallest value of A?

14.   In the correctly worked out addition problem below, each letter represents a different digit. How many possible values are there for the digit C?

ADF
MNP
———— +
CCCC

15. If a semi-circle is inscribed in a rectangle, what can be the ratio between the width and the length of the rectangle?

16. A jar contains only blue, green, and white marbles. If a marble is selected at random from the jar, the probability that it is not blue is $\dfrac{6}{11}$ and the probability that it is not white is $\dfrac{9}{11}$. What is the probability that it is green?

17.   If $S = 5 + 10 + 15 + 20 + \cdots + 85$ and $T = 6 + 11 + 16 + \cdots + 86$, what is the value of $|T - S|$?

18.   A small business employs 58 people. One day, six times more women than men were absent from work due to a flu epidemic. As a result, there were five times more men at work than women. How many more of the employees are men than women?

39.1 Self-grade Diagnostic Eight

Question	1	2	3	4	5	6	7	8
Level	1	1	1	1	2	2	3	3
Topic	N	N	G	A	N	A	S	G
Answer	B	A	E	A	E	D	E	D
Correct								
Incorrect								
Skipped								

Question	9	10	11	12	13	14	15	16	17	18
Level	2	2	2	3	4	3	4	4	3	4
Topic	G	A	S	N	A	N	G	S	N	N
Answer	120	13	400	6.4	14	1	1/2	4/11	17	0
Correct										
Incorrect										
Skipped										

Are your incorrect answers:

mostly on questions in the first part of the test?	Yes	No
mostly on questions that have difficulty levels 3 and 4?	Yes	No
mostly from a specific topic (geometry, algebra, arithmetic)?	Yes	No
somewhat evenly spread throughout the test?	Yes	No
mostly due to minor errors?	Yes	No
mostly due to numeric errors?	Yes	No

39.2 Diagnostic Eight Solutions

Question 1

Since:
$$k^{-1} = \frac{1}{k}$$

we have:
$$a^{-1}/b^{-1} = \frac{1}{a} \div \frac{1}{b} = \frac{b}{a}$$

Substituting the values, we obtain:
$$\frac{2}{-3} = -\frac{2}{3}$$

Question 2

Execute the operations from left to right, in order:
$$4\nabla 2 = 2^2 - 4^2 = 4 - 16 = -12$$

$$-12\nabla 1 = 1^2 - (-12)^2 = 1 - 144 = -143$$

Question 3

Since there is a multiple of 3 every three consecutive numbers, the number 188 will be large enough to fit 62 multiples of 3:
$$\lfloor \frac{188}{3} \rfloor = 62$$

Question 4

Since $y = ax + t$ is perpendicular to $x = by + m$, rewrite the second line in slope-intercept form:

$$
\begin{aligned}
by &= x - m \\
y &= \frac{x - m}{b} \\
y &= \frac{1}{b}x - \frac{m}{b}
\end{aligned}
$$

Then the slopes of the two lines must have a product equal to -1:

$$
\frac{1}{b} \times a = -1
$$

and, therefore:

$$
a = -b
$$

Question 5

The total number of points obtained for the 6 tests is:

$$
85 \times 6 = 510
$$

Before the last test, the cumulated number of points was:

$$
510 - 90 = 420
$$

The point average before the last test was:

$$
\frac{420}{5} = 84
$$

Question 6

The function $f(x)$ divides x by its absolute value. If $x > 0$, $f(x) = 1$ and if $x < 0, f(x) = -1$. Therefore, the only choice that is not true for any x is $f(x + 1) = f(x - 1)$. Indeed, if $x = 0.5$:

$$
\begin{aligned}
f(1.5) &= 1 \\
f(-0.5) &= -1
\end{aligned}
$$

Question 7

For each of the four semaphores there are three choices, for a total number of combinations:

$$3 \cdot 3 \cdot 3 \cdot 3 \cdot = 3^4 = 81$$

Question 8

Denote the intersection of AC and MN with P. It is easy to prove (SAS) that $\triangle AMP \equiv \triangle PNC$. The area of the parallelogram $BCNM$ is equal to the area of $\triangle ABC$. The area of $\triangle ABC$ is:

$$\frac{1}{2} \cdot \sqrt{3} \cdot \frac{3}{2} = \frac{3\sqrt{3}}{4}$$

Question 9

The angles a and c can be written as:

$$a = b \cdot 1.35$$
$$c = b \cdot 0.65$$

Since the sum of all three angles is $360°$:

$$1.35b + b + 0.65b = 3b = 360°$$

we have that $b = 120°$.

Question 10

The inequality $|x - 2| < 7$ is equivalent to:

$$-7 < x - 2 < 7$$

Or

$$-5 < x < 9$$

There are $9 - (-5) - 1 = 13$ integer values for x that satisfy the inequality.

Question 11

Refer to the following figure:

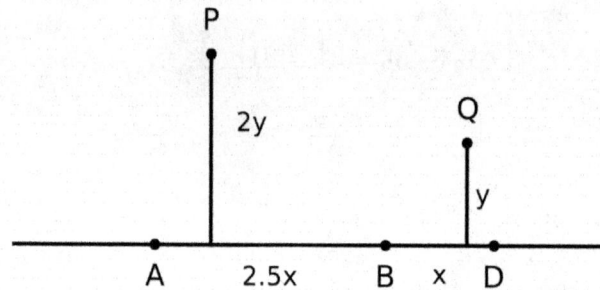

If $DeltaBDQ$ has area:

$$\frac{xy}{2} = \frac{xy}{2}$$

$\triangle ABP$ has area:

$$\frac{(2.5x)(2y)}{2} = \frac{5xy}{2}$$

Since the area of $\triangle ABP$ is 5 times as large as the area of $\triangle BDQ$, then it is 400% larger.

Question 12

If N loaves were ordered, the bakery baked $1.12N$ loaves. If 5% of the total number of loaves were rejected, the remaining amount of loaves was:

$$1.12 \cdot N \cdot 0.095 = 1.064N$$

Therefore, there were still 6.4% more loaves than ordered.

Question 13

We have:

$$-4 < x - 3 < 4$$

$$-1 < x < 7$$

Add 7:

$$6 < x + 7 < 14$$

If the inequality above is true, then we also have:

$$-14 < x + 7 < 14$$

and
$$|x + 7| < 14$$

Therefore, $A = 14$.

Question 14

Since, by adding two digits, the largest possible carry is 1, C can only have the value 1. We have to find an example that shows it is possible to obtain the number 1111 by adding two 3-digit numbers with all different digits. Such an example is easily found:

$$1111 = 235 + 876$$

Therefore, there is only one value of C that satisfies.

Question 15

If the semi-circle is inscribed in a rectangle, then the circle is tangent to three sides of the rectangle, as in the figure:

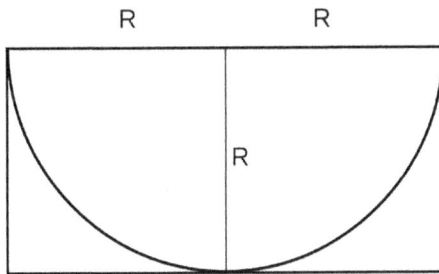

The width of the rectangle is equal to the radius of the circle and the length of the rectangle is equal to the diameter of the circle. The ratio between the width and the length is 1/2.

Question 16

Denote the probability to be of color K by $P(K)$. If the marble is not blue, then it is green or white. The probability for it to be green or white is:

$$P(G) + P(W) = \frac{6}{11}$$

The probability that it is not white is the same as the probability that it is blue or green:

$$P(B) + P(G) = \frac{9}{11}$$

Adding these two equations:

$$P(G) + P(W) + P(G) + P(B) = \frac{15}{11}$$

and applying the fact that the probability of getting a marble of any color is 1 (certain event):

$$P(G) + P(W) + P(B) = 1$$

we obtain:

$$P(G) + 1 = \frac{15}{11}$$

and

$$P(G) = \frac{4}{11}$$

Question 17

Since:

$$T - S = 6 - 5 + 11 - 10 + \cdots 86 - 85$$

Each pair of terms counting from the left have a difference of 1. It is sufficient to count the total number of terms:

$$\frac{85}{5} = 17$$

$|T - S| = 17$

Question 18

The table summarizes the number of people at work and the number of people that are absent:

Absent		At work	
Women	Men	Women	Men
6k	k	p	5p

The total number of people in the business is $6k+k+p+5p = 7k+6p$.

Since $7k + 6p = 58$ and the number of people cannot be fractional, we attempt to find values for p and k that match. Since 58 is even and $6p$ is also even, $7k$ must be even. Therefore we must try to use even multiples of k such as: $14, 28, 42$. Larger numbers would exceed the total. Subtract each of these from 58 to find which difference is a multiple of 6 (if any):

$$58 - 14 = 44$$

$$58 - 28 = 30$$

$$58 - 42 = 16$$

Of these, only 30 is a multiple of 6. Using $p = 5$ and $k = 4$ we can now put actual values in the table:

Absent		At work	
Women	Men	Women	Men
24	4	5	25

There are $4 + 25 = 29$ men and $24 + 5 = 29$ women. The difference is zero.

ALGEBRA QUIZ SEVEN

What is the largest value of k if:

1. 2^k is a factor of 648

2. 3^{k-1} is a factor of 648

3. 8^k is a factor of 24^{10}

4. 6^k is a factor of 24^{10}

5. 5^k is a factor of 175^8

6. 16^k is a factor of 8^{20}

Solve for x in each of the equations:

7. $3^{k-1} + 3^k + 3^{k+1} = 117$

8. $\left(\dfrac{5}{8}\right)^x \cdot \sqrt[x]{\dfrac{64}{25}} = \left(\dfrac{5}{8}\right)^3$

9. $\sqrt[k]{25^3} \cdot 5^k = 5^5$

10. $2^x \cdot \left(\frac{1}{8}\right)^{3x} = 16^{10}$

Answers to questions 1-6

1. $648 = 2^3 \cdot 3^4, \quad k = 3$

2. $648 = 2^3 \cdot 3^4, \quad k - 1 = 4, \quad \rightarrow k = 5$

3. $24^{10} = 8^{10} \cdot 3^{10}, \quad k = 10$

4. $24^{10} = 6^{10} \cdot 4^{10}, \quad k = 10$

5. $175^8 = 7^8 \cdot (5^2)^8 = 7^8 \cdot 5^{16}, \quad k = 16$

6. $8^{20} = (2^3)^{20} = 2^{60}, \quad 16^k = 2^{4k}, \quad 4k = 60 \rightarrow k = 15$

Answers to questions 7-10

7. $3^{k-1} + 3^k + 3^{k+1} = 3^{k-1}(1 + 3 + 9) = 13 \cdot 3^{k-1} = 13 \cdot 3^2, \quad k - 1 = 2, \rightarrow k = 3$

8. $\left(\dfrac{5}{8}\right)^x \cdot \sqrt[x]{\dfrac{64}{25}} = \left(\dfrac{5}{8}\right)^x \cdot \left(\dfrac{8}{5}\right)^{\frac{2}{x}} = \left(\dfrac{5}{8}\right)^x \cdot \left(\dfrac{5}{8}\right)^{-\frac{1}{x}} = \left(\dfrac{5}{8}\right)^{x - \frac{2}{x}}$

$x - \dfrac{2}{x} = 3, \quad x^2 - 3x - 2 = 0, \quad (x - 2)(x - 1) = 0, \rightarrow x = 2$

Note that 1 cannot be the index of a radical, hence $x = 1$ is not a solution.

9. $\sqrt[k]{(5^2)^3} \cdot 5^k = 5^{\frac{6}{k}} \cdot 5^k = 5^{\frac{6}{k}+k} = 5^5$

$\dfrac{6}{k} + k = 5, \quad k^2 - 5k + 6 = 0, \quad (k - 2)(k - 3) = 0 \quad \rightarrow k = 2 \text{ or } k = 3$

10. $2^x \cdot \dfrac{1}{(2^3)^{3x}} = 2^x \cdot \dfrac{1}{2^{9x}} = 2^{x-9x} = 2^{-8x}$

$16^{10} = (2^4)^{10} = 2^{40}, \quad \rightarrow -8x = 40, \quad x = -5$

1.		$AQ = 6$ $QC = 4$	$AP \cdot AB = k$	$k =$
2.		$AB = 15$ $PC = 12$	$\dfrac{QB}{BC} = k$	$k =$
3.		DB bisects $\angle ADC$ $BD = 5$	$CD \cdot AD = k$	$k =$
4.		$ABCD$ is square $FM = 7$ $PQ = 9$	$FE \cdot MQ = k$	$k =$

Answers to questions 1-4

1. Similar triangles $\triangle APQ \equiv \triangle ACB$:

$$\frac{AQ}{AB} = \frac{AP}{AC}$$

$$\frac{6}{AB} = \frac{AP}{10}$$

$$AB \cdot AP = 60, \quad \rightarrow k = 60$$

2. Similar triangles $\triangle BQA \equiv \triangle CPB$:

$$\frac{AB}{PC} = \frac{QB}{BC}$$

$$\frac{AB}{PC} = \frac{15}{12} = \frac{5}{4}$$

$$\frac{QB}{BC} = \frac{5}{4}, \quad \rightarrow k = 1.25$$

3. Similar triangles $\triangle ADB \equiv \triangle BDC$:

$$\frac{AD}{BD} = \frac{BD}{CD}$$

$$AD \cdot CD = BD^2 = 25, \quad \rightarrow k = 25$$

4. Since $ABCD$ is a square, $FE \| PQ$ and there are similar triangles $\triangle MFE \equiv \triangle MQP$:

$$\frac{FM}{FE} = \frac{MQ}{PQ}$$

$$\frac{7}{FE} = \frac{MQ}{9}$$

$$FE \cdot MQ = 63, \quad \rightarrow k = 63$$

	Simplify		Simplify
1.	$\frac{\frac{2}{3}}{8} =$	**2.**	$\frac{\frac{5}{15}}{9} =$
3.	$\frac{1}{\frac{1}{5}} =$	**4.**	$\frac{\frac{3}{8}}{\frac{12}{16}} =$
5.	$\frac{\frac{ab}{ac}}{\frac{bc}{ab}} =$	**6.**	$\frac{ab-ac}{ma-na} =$
7.	$\frac{\frac{ma}{nb}}{\frac{na}{mc}} =$	**8.**	$\frac{1}{\frac{1}{3}} =$
9.	$\frac{2^{-1}}{2^{-3}} =$	**10.**	$\frac{2}{\frac{a}{m}} =$
11.	$\frac{1^{-1}}{(-1)^1} =$	**12.**	$\frac{0^2}{(-1)^0} =$
13.	$\frac{\frac{-1}{-1}}{(-1)^{-1}} =$	**14.**	$\frac{1-x}{x-1}$
15.	$\frac{ab-ac}{dc-db} =$	**16.**	$\frac{12!}{20!} \cdot \frac{19!}{11!} =$

Answers to questions 1-10

	Simplify		Simplify
1.	$\frac{\frac{2}{3}}{8} = \frac{2}{3} \cdot \frac{1}{8} = \frac{1}{12}$	**2.**	$\frac{5}{\frac{75}{9}} = \frac{5}{1} \cdot \frac{9}{75} = \frac{3}{5}$
3.	$\frac{1}{\frac{1}{5}} = 5$	**4.**	$\frac{\frac{3}{8}}{\frac{12}{16}} = \frac{3}{8} \cdot \frac{16}{12} = \frac{1}{2}$
5.	$\frac{\frac{ab}{ac}}{\frac{bc}{ab}} = \frac{ab}{ac} \cdot \frac{ab}{bc} = \frac{ab}{c^2}$	**6.**	$\frac{ab-ac}{ma-na} = \frac{a(b-c)}{a(m-n)} = \frac{b-c}{m-n}$
7.	$\frac{\frac{ma}{nb}}{\frac{na}{mc}} = \frac{ma}{nb} \cdot \frac{mc}{na} = \frac{m^2c}{n^2b}$	**8.**	$\frac{1}{\frac{1}{3}} = 3$
9.	$\frac{2^{-1}}{2^{-3}} = \frac{1}{2} \cdot \frac{8}{1} = 4$	**10.**	$\frac{2}{\frac{a}{m}} = \frac{2}{1} \cdot \frac{m}{a} = \frac{2m}{a}$
11.	$\frac{1^{-1}}{(-1)^1} = \frac{1}{-1} = -1$	**12.**	$\frac{0^2}{(-1)^0} = \frac{0}{1} = 0$
13.	$\frac{\frac{-1}{-1}}{(-1)^{-1}} = \frac{1}{-1} = -1$	**14.**	$\frac{1-x}{x-1} = -\frac{x-1}{x-1} = -1$
15.	$\frac{ab-ac}{dc-db} = \frac{a(b-c)}{d(c-b)} = -\frac{a}{d}$	**16.**	$\frac{12!}{20!} \cdot \frac{19!}{11!} = \frac{11! \cdot 12}{19! \cdot 20} \cdot \frac{19!}{11!} = \frac{3}{5}$

DATA ANALYSIS QUIZ SEVEN

1. Throw 4 dice. What is the probability of obtaining at least one die that shows 1?

2. How many 2-digit numbers can be written using only the digits 3, 4, and 0?

3. What is the probability of the event that is complementary to the event "by rolling a die we will get a 2 or a 5?"

4. Let A be an event that you roll a die and obtain one of 1, 2, and 3, and B be an event that you roll a die and obtain 3, 4, or 5. If the probability of A is p, what is the probability of B?

5. Let A be the event that you toss a coin and obtain heads, and B be the event that you toss a coin and obtain tails. These events are (check all that apply):

- complementary
- independent
- conditional
- mutually exclusive

6. At a car dealership, the probability that a blue sedan is sold within the next hour is 0.39, and the probability that a sedan is sold within the next hour is 0.65. What is the probability a blue car is sold, given that it is a sedan?

7. In the city of Xorg, the probability that an elderly man will arrive on horseback within the next day is 0.34, and the probability that someone will arrive on horseback is 0.51. What is the probability that an elderly man will arrive, given that the person is on horseback?

Answers to questions 1-7

1. This event is the opposite of not having any die that shows 1. The probability to not have any die that shows 1 is:

$$\left(\frac{5}{6}\right)^4$$

The probability for the required event is $1 - \left(\frac{5}{6}\right)^4$.

2. 0 cannot be the leftmost digit, if the number has to be a 2-digit number. The leftmost digit can have 2 different values (3 and 4) and the rightmost digit can have any of the three values. The total number of possibilities is $2 \times 3 = 6$.

3. The probability to get a 2 or a 5 is:

$$\frac{1}{6} + \frac{1}{6} = \frac{2}{6} = \frac{1}{3}$$

The probability of the complementary event is $1 - \frac{1}{3} = \frac{2}{3}$.

4. p as well.

5. The events are complementary, independent, and mutually exclusive.

6. The events are conditional. The probability for a blue car to be sold in the next hour, provided that it is a sedan, is:

$$P(\text{blue}|\text{sedan}) = \frac{P(\text{blue sedan})}{P(\text{sedan})} = \frac{0.39}{0.65} = \frac{3 \cdot 13}{5 \cdot 13} = \frac{3}{5} = 0.6$$

7. The events are conditional. The probability for an elderly man to be the rider of the horse is:

$$P(\text{elderly}|\text{on horseback}) = \frac{P(\text{elderly on horseback})}{P(\text{ on horseback})} = \frac{0.34}{0.51} = \frac{2 \times 17}{3 \times 17} = \frac{2}{3}$$

Diagnostic Nine

This math section contains 18 questions to be solved in 25 minutes.

You are allowed to use a calculator. However, in order to minimize the time needed to complete the test, it is important to use the calculator as little as possible! Keep in mind:

- Entering operations in the calculator is more time consuming than performing the operations mentally.

- Data entry errors will be made in addition to other errors. Entering data is, in itself, a possible cause for error.

You can reduce the use of the calculator by memorizing well a short list of commonly used numbers: perfect squares from 1 to 20^2, powers of 2 from 2^0 to 2^{10}, some frequently used Pythagorean triples, etc. The list can be found in Appendix A, as well as on www.mathinee.com.

As you solve problems, build on the ability to recognize the numbers in this list. Every time you encounter them, tell yourself "hey, this is a power of 2", or "hey, this is a Pythagorean triple." After a short while, you will have memorized the list almost completely. But going over the list 'cold' a few times is also helpful, so go ahead and open Appendix A!

Before starting to solve, spend a few seconds to think about focusing, about making sure that you are ready to pay attention to every detail.

1. What is 400% of $\frac{1}{p}$?

(A) $\dfrac{4}{p}$

(B) $4p$

(C) $\dfrac{400}{p}$

(D) $400p$

(E) $\dfrac{400}{p}$

(A) (B) (C) (D) (E)

---

2. If a number is 87.5% of another number, what could be the ratio of the two numbers?

(A) $35:4$

(B) $7:1$

(C) $7:5$

(D) $8:7$

(E) $8:15$

(A) (B) (C) (D) (E)

3. Two perpendicular lines l and m intersect at point P, inside a circle with center O. With the notation from the figure, what is the sum of the measures of the arcs $\overset{\frown}{BC}$ and $\overset{\frown}{AD}$, in degrees of arc?

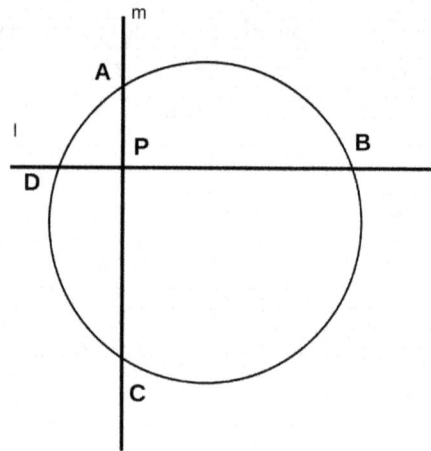

(A) $90°$

(B) $120°$

(C) $150°$

(D) $180°$

(E) $210°$

(A) (B) (C) (D) (E)

4. Find the value of x:

$$5^{10-x} = 15625$$

(A) 4

(B) 5

(C) 6

(D) 7

(E) 8

6. If $f(x) = 2x + 3$ and $g(f(x)) = 4x$, which of the following choices is $g(x)$?

(A) $2x - 3$

(B) $\dfrac{1}{2}x + 3$

(C) $2x + 6$

(D) $2x - 6$

(E) $\dfrac{1}{2}x - 3$

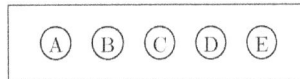

5. 15 objects were purchased for a tombola. The total cost was \$860. Prizes consist of 5 objects placed in a bag and each object may cost either \$100, \$70 or \$50. There are three prizes, each of different value. What is the maximum value the largest prize could have?

(A) 210

(B) 287

(C) 340

(D) 550

(E) 740

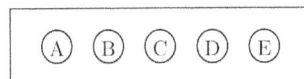

7. The graph shows the number of pairs of beach shoes an online store projected will be sold vs. the number of pairs that were actually sold. The projected number is shaded grey and the actual sales numbers are black bars. In what month was there the largest percent difference between the projected sales and the actual sales?

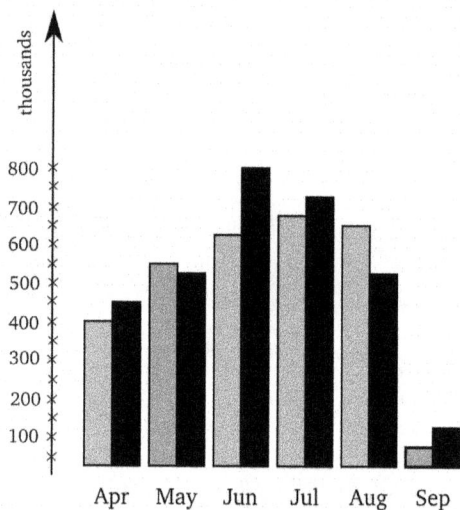

(A) May

(B) June

(C) July

(D) August

(E) September

$$\boxed{\text{(A)} \quad \text{(B)} \quad \text{(C)} \quad \text{(D)} \quad \text{(E)}}$$

8. A cube with a diagonal of length $\sqrt{5}$ has volume:

(A) $15\sqrt{5}$

(B) $\dfrac{5\sqrt{5}}{3\sqrt{3}}$

(C) $\dfrac{5\sqrt{5}}{2\sqrt{2}}$

(D) 5

(E) $5\sqrt{5}$

$$\boxed{\text{(A)} \quad \text{(B)} \quad \text{(C)} \quad \text{(D)} \quad \text{(E)}}$$

9. Which of the choices is equivalent to $5^a \cdot 3^b$?

(A) 15^{a+b}

(B) 8^{ab}

(C) 8^{a+b}

(D) 15^{ab}

(E) $15^a \cdot 3^{b-a}$

$$\boxed{\text{(A)} \quad \text{(B)} \quad \text{(C)} \quad \text{(D)} \quad \text{(E)}}$$

10. A rectangular box placed on a level surface has base dimensions $4, 5$ and height 15. The liquid inside the box reaches a level of 6 units. If we place the box to have the base with dimensions $4, 15$ approximately how many units from the top of the box will the level of the liquid be?

(A) 2
(B) 3
(C) 5
(D) 6
(E) 10

Ⓐ Ⓑ Ⓒ Ⓓ Ⓔ

11. In the figure, what is the ratio between the shaded area and the area of the triangle ABC?

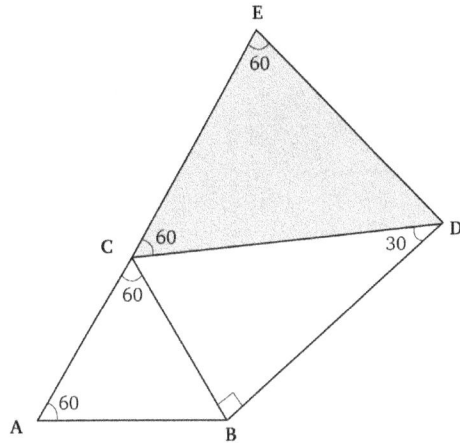

(A) $2 : 1$
(B) $3 : 2$
(C) $5 : 2$
(D) $4 : 1$
(E) $\sqrt{3} : 1$

Ⓐ Ⓑ Ⓒ Ⓓ Ⓔ

12. The speed of a vehicle varied with time as in the figure. What was the average speed, in miles per hour, of the vehicle during the 6 hours represented by the graph?

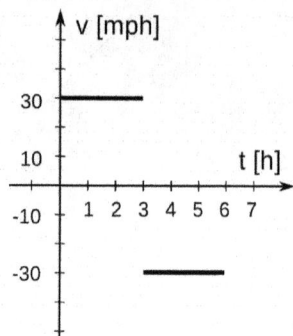

(A) -30

(B) -15

(C) 0

(D) 15

(E) 30

Ⓐ Ⓑ Ⓒ Ⓓ Ⓔ

13. If ♯ is an operation that represents choosing the number with the larger sum of the digits, and ♭ represents choosing the number with the smaller sum of the digits, what is (20♯19)♭27 ?

(A) 9

(B) 10

(C) 19

(D) 20

(E) 27

Ⓐ Ⓑ Ⓒ Ⓓ Ⓔ

14. If $5^{k-1} = p\left(5^k + 5^{k+1}\right)$, then the value of p must be:

(A) $\dfrac{1}{5}$

(B) $\dfrac{2}{5}$

(C) $\dfrac{1}{15}$

(D) $\dfrac{1}{30}$

(E) 30

Ⓐ Ⓑ Ⓒ Ⓓ Ⓔ

15. If d dollars and c cents can pay for a apples, what amount, in cents, is needed to pay for k apples?

(A) $(d + 100c)\dfrac{a}{k}$

(B) $\dfrac{k}{a}(100d + c)$

(C) $\dfrac{d + c}{ak}$

(D) $\dfrac{d + c}{a}k$

(E) $(d + 100c)\dfrac{k}{a}$

Ⓐ Ⓑ Ⓒ Ⓓ Ⓔ

16. A sequence starts with the terms $1, 2, 3$. Each succesive term is generated so as to increase the average of the terms (up to and including itself) by 2. What is the 5th term of the sequence?

(A) 5

(B) 7

(C) 10

(D) 14

(E) 24

Ⓐ Ⓑ Ⓒ Ⓓ Ⓔ

44.1   Self-grade Diagnostic Nine

Question	1	2	3	4	5	6	7	8
Level	1	1	1	2	2	2	2	3
Topic	N	N	G	A	N	A	S	G
Answer	A	D	D	A	C	D	E	B
Correct								
Incorrect								
Skipped								

Question	9	10	11	12	13	14	15	16
Level	2	3	3	3	4	4	3	4
Topic	N	G	G	N	N	A	N	S
Answer	E	A	D	E	E	D	B	D
Correct								
Incorrect								
Skipped								

Are your incorrect answers:

mostly on questions in the first part of the test?	Yes	No
mostly on questions that have difficulty levels 3 and 4?	Yes	No
mostly from a specific topic (geometry, algebra, arithmetic)?	Yes	No
somewhat evenly spread throughout the test?	Yes	No
mostly due to minor errors?	Yes	No
mostly due to numeric errors?	Yes	No

A-algebra, N-number sense, G-geometry, S-statistics

44.2 Diagnostic Nine Solutions

Question 1

Use the definition of percentages:

$$\frac{400}{100} \times \frac{1}{p} = \frac{4}{p}$$

Question 2

Denote one of the numbers by N, the other number M is:

$$M = \frac{87.5}{100} \times N = \frac{875}{1000} \times N = \frac{5^3 \times 7}{5^3 \times 8}N = \frac{7}{8}N$$

Then we have:

$$N = \frac{8}{7}M$$

and

$$M = \frac{7}{8}N$$

Only one of the ratios is an answer choice.

Question 3

Due to symmetry with respect to an axis that makes $45°$ angles with both lines, the arcs $\overset{\frown}{AB}$ and $\overset{\frown}{CD}$ are congruent:

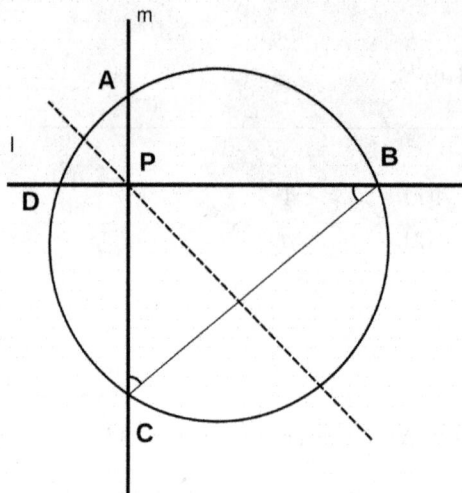

The angles $\angle BDC$ and $\angle ACB$ are both equal to half of the arc $\overset{\frown}{AB}$:

$$\angle BDC = \angle ACB = \frac{1}{2}\overset{\frown}{CD} = \frac{1}{2}\overset{\frown}{AB}$$

Since $\angle CPB = 90°$, we have that $\angle PBC = \angle ACB = 45°$.

Therefore, the arcs intercepted by these two angles have measure $90°$ and their sum is $180°$. The remaining two arcs must have measures that have a sum of $180°$ also:

$$\overset{\frown}{AD} + \overset{\frown}{BC} = 180°$$

Question 4

Write 15625 as a power of 5:

$$15625 = 15000 + 625 = 15 \times 5^3 \times 8 + 5^4 = 5^4 \times 24 + 5^4 = 5^4 \times (24+1) = 5^6$$

Therefore, $10 - x = 6$ and $x = 4$

Question 5

The smallest possible prize value is $50 \times 5 = 250$ dollars. If we make the smallest prize exactly equal to 250, the values of the two larger prizes must total $860 - 250 = 510$. Since $70 \times 3 = 210$, there must be three objects valued at 70 among the first two prizes. The remaining objects can be multiples of 50 that total 300 dollars. Since only $10 - 3 = 7$ more objects are needed, only one of them must cost 100 dollars. The prizes are as follows:

$$100 + 70 + 70 + 50 + 50 \;=\; 340$$

$$70 + 50 + 50 + 50 + 50 \;=\; 270$$

$$50 + 50 + 50 + 50 + 50 \;=\; 250$$

The largest possible value for the first prize is 340 dollars.

Question 6

From

$$g(f(x)) = 4x$$

we infer that:

$$
\begin{aligned}
g(2x + 3) &= 4x \\
g(2x + 3) &= 2(2x + 3) - 6 \\
g(x) &= 2x - 6
\end{aligned}
$$

Question 7

The percent difference is:

$$\frac{|\text{actual} - \text{projected}|}{\text{projected}}$$

If we really want to compute the percentages, we have for each month:

Month	Actual sales	Projected sales	% difference
Apr	450	400	$\frac{50}{450} \times 100\% \approx 11\%$
May	525	550	$\frac{25}{550} \times 100\% \approx 5\%$
June	800	650	$\frac{150}{650} \times 100\% \approx 23\%$
July	700	650	$\frac{50}{650} \times 100\% \approx 77\%$
August	500	650	$\frac{150}{650} \times 100\% \approx 23\%$
September	100	50	$\frac{50}{50} \times 100\% \approx 100\%$

However, it is easy to see that the actual sales were twice the projected sales in September, while for any other month the factor between the actual and the projected sales was much smaller. No computation is actually needed.

Question 8

If the side of the cube is equal to s, then the diagonal has length:

$$d = \sqrt{s^2 + s^2 + s^2} = s\sqrt{3}$$

Therefore:

$$s\sqrt{3} = \sqrt{5}$$

The volume of the cube is s^3:

$$V = s^3 = \left(\frac{\sqrt{5}}{\sqrt{3}}\right)^3 = \frac{5\sqrt{5}}{3\sqrt{3}}$$

Question 9

Since the bases are different, the expression cannot be simplified. We simply have to determine which of the choices is equivalent to it. Of all choices, we notice that the last one is a choice in which a factors of 3 have been incorporated into the first power:

$$
\begin{aligned}
15^a \cdot 3^{b-a} &= 5^a \cdot 3^a \cdot 3^{b-a} \\
&= 5^a \cdot 3^{a+b-a} \\
&= 5^a \cdot 3^b
\end{aligned}
$$

Question 10

The volume of the liquid remains the same:

$$6 \times 4 \times 5 = 4 \times 15 \times h$$

We find that $h = 2$ units.

Question 11

By chasing angles in the figure, we find that the triangle CBD is a $30 - 60 - 90$ triangle. Therefore, the side of the shaded triangle is twice as long as the side of $\triangle ABC$. Since both the shaded triangle and $\triangle ABC$ are equilateral, they are similar. If the side lengths are in a ratio of $2 : 1$ then the areas are in a ratio of $4 : 1$.

Question 12

To obtain the average speed, divide the total distance by the total time. The total distance is:

$$30 \times 3 + 30 \times 3 = 180$$

The total time is 6 units. The average velocity is:

$$\frac{180}{6} = 30$$

Question 13

The digit sums of the numbers involved are: $S(20) = 2, S(19) = 10$, and $S(27) = 9$.

Processing the expression from left to right, apply the definition of $\sharp$:

$$10 \sharp 19 = 19$$

and the definition of $\flat$:

$$19 \flat 27 = 27$$

Question 14

Use the properties of exponents and factor out a 5^k:

$$5^k + 5^{k+1} = 5^k + 5^k \cdot 5 = 5^k(1 + 5) = 6 \cdot 5^k$$

Then, solve for p:

$$5^{k-1} = p \cdot 6 \cdot 5^k$$

Divide by 5^{k-1}:

$$5^{k-1} = p \cdot 6 \cdot 5^k$$

$$1 = p \cdot 6 \cdot 5$$

$$1 = 30p$$

$$p = \frac{1}{30}$$

Question 15

Convert the amount to cents: $100d + c$ and calculate the price of 1 apple:

$$\frac{100d + c}{a}$$

Multiply this price with the number of apples (k):

$$\frac{100d + c}{a} \cdot k$$

Question 16

The average of the first three terms is 2. The 4th term is such that the average of the first 4 terms is 4:

$$\frac{1 + 2 + 3 + x}{4} = 4$$

We solve for x and find that $x = 10$. The 5th term is such that the average of the first 5 terms is 6:

$$\frac{1 + 2 + 3 + 10 + y}{5} = 6$$

We find $y = 14$.

APPENDIX A

The appendix contains facts that should be memorized in order to become faster at solving SAT problems.

Number and operation facts:

- 1 is not prime, nor composite - is is an improper prime

- 2 is the only even prime number

- 0 is not positive, nor negative

- 0 is even

- 0 is a multiple of any number

- division by 0 is undefined

Rules of parity in operations with integers:

- even + even = even
- odd + odd = even
- even + odd = odd
- even $\times$ any integer = even
- odd $\times$ odd = odd

The first 11 powers of 2:

$$2^0 = 1$$
$$2^1 = 2$$
$$2^2 = 4$$
$$2^3 = 8$$
$$2^4 = 16$$
$$2^5 = 32$$
$$2^6 = 64$$
$$2^7 = 128$$
$$2^8 = 256$$
$$2^9 = 512$$
$$2^{10} = 1024$$

The prime numbers smaller than 110:

$$2, 3, 5, 7, 11, 13, 17, 19, 23, 29, 31, 37, 41, 43$$

$$47, 53, 59, 61, 67, 71, 73, 79, 83, 89, 97, 101, 109$$

The number of prime numbers smaller than 100: there are 25 primes smaller than 100.

The perfect squares of the numbers from 1 to 20:

$$1, 4, 9, 16, 25, 36, 49, 64, 81, 100, 121$$

$$144, 169, 196, 225, 256, 289, 324, 361, 400$$

Some perfect cubes:

$$1^3 = 1, 2^3 = 8, 3^3 = 27, 4^3 = 64, 5^3 = 125, 6^3 = 216, 8^3 = 512$$

The factorials from 1 to 6:

$$
\begin{aligned}
0! &= 1 \\
1! &= 1 \\
2! &= 2 \\
3! &= 6 \\
4! &= 24 \\
5! &= 120 \\
6! &= 720
\end{aligned}
$$

Pythagorean triples:

$$3, 4, 5$$

$$5, 12, 13$$

$$7, 24, 25$$

$$8, 15, 17$$

The multiples of these triples are also Pythagorean.

The approximations of some irrational numbers:

$$\sqrt{2} \approx 1.41$$
$$\sqrt{3} \approx 1.73$$
$$\frac{\sqrt{2}}{2} \approx 0.71$$

www.ingramcontent.com/pod-product-compliance
Lightning Source LLC
Chambersburg PA
CBHW051210200326
41519CB00025B/7069